ONE OF US

PART 3 OF THE VILLAGE TRILOGY

RACHEL MCLEAN

This is a work of fiction. Names, characters, businesses, places, events and incidents are either the products of the author's imagination or used in a fictitious manner. Any resemblance to actual persons, living or dead, or actual events is purely coincidental.

Catawampus Press

catawampus-press.com

JOIN MY BOOK CLUB

If you want to find out more about the characters in *One of Us* - how they escaped the floods and arrived at the village - you can read the prequel for free by joining my book club.

Go to rachelmclean.com/underwater to read *Underwater* for free.

Happy reading!

Rachel McLean

1

"YOU'VE *WHAT*?"

Leah Golder stared at her son across the table. Beside him, Jess squirmed in her seat. She was a grown woman. She'd dealt with much worse than this; so why was she feeling scared?

"You heard, Mum," Zack replied. "I thought you'd be pleased."

Leah's gaze shifted from Zack to Jess. Jess gave her a nervous smile. In an instant, the woman had gone from beaming and friendly to shocked and hostile.

Jess looked down at the table between them. On it was a cake and a plate of biscuits. Leah had probably used a week's rations for this meal.

Next to her mother, Zack's youngest sister – was it Shirley or Rose? Jess struggled to remember – shovelled cake into her mouth, knowing she wouldn't get this opportunity again in a hurry. The girl was eleven years old, over ten years younger than her brother and his twin, Sam. Twenty years younger than Jess. Sam was currently stuffing the remains of a fish pie into his mouth. Crumbs slipped from his lips and rained onto the tablecloth. Jess wondered how often it got used.

"Well, I knew you were… close," said Leah, her stare moving back to Zack. "But I never thought it would come to…"

She put down her knife and picked up a napkin. She wiped her lips, her eyes closed.

Zack looked at his dad, Tim. Jess followed his gaze, hoping for some help.

Tim cleared his throat. "Don't get me wrong, Jess. This is nothing personal. And it's not because you're the steward, or anything like that. It's just – well – you're… it's the age difference."

"Don't be daft, Dad," said Zack. "Who cares about nine years? I love her."

He grabbed her hand under the table. She thought of the ring he'd slipped on her finger the previous night. Plastic, scavenged from a landfill site near his work, but it was the thought that counted. She recalled the rings she'd given to her sister-in-law Ruth six months ago, after Ruth had been kidnapped and Jess and Zack had gone to save her. Her mother Sonia's rings, kept safe in a tin for the previous five years. *I don't need them*, she'd said. She'd meant it at the time.

She took a deep breath and smiled at her prospective mother-in-law. "I'm sorry to have surprised you, Leah." She couldn't bring herself to call her Mrs Golder. Leah and Tim had been Leah and Tim before she'd got to know Zack. Maybe that was the problem.

"You taught him," Leah replied. "You were his teacher." She picked up the teapot and stared into it as if it would explain everything.

"For one year. Six years ago. And it's not as if our relationship started then."

"Hmm," said Tim. "Maybe she's got a... what was that?"

Jess felt a tremor ripple through her. She tightened her grip on Zack's hand and stood up.

The room went quiet. Rose-or-Shirley dropped her fork and squealed.

Leah put down the teapot. It was brown, chipped on the spout, its shine lost except for one patch on the side that reflected the light from the candles on the windowsill. It was almost completely dark now, the darkness exaggerated by the fact that lights-out was just fifteen minutes ago.

Maybe Jess had imagined it.

"Did you feel that?" she said.

"Yes," said Zack. He dropped her hand and went to the window. "Shit."

Jess ignored Leah's frown. "What?" She slid in behind him at the window, peering over his shoulder.

There it was again. Lower, this time, and less abrupt. A rumbling sound, and a low tremor running through them.

She turned to Zack's family. "I've got to go. Sorry."

She dashed towards the front door and yanked it open. The village was in darkness. Opposite, a door opened, the faint glow of a candle silhouetting a man, raising himself up to get a better view.

She didn't pause to check if Zack was behind her, but instead ran towards the centre of the village. As she approached the community centre, there was another bang, two beats and then a splintering sound. It was coming from the north.

She almost skidded to turn along the road that led to the northern edge of the village, wondering if any of the other council members had heard it. If they would be out too.

Doors were opening as she hurried past. People

muttered as they stumbled out into darkness. Ahead of her, slowly gaining in intensity, a glow rose above the rooftops.

She felt her stomach lurch. *Oh no, please no*. Hadn't they been through enough? Her own sister-in-law had been kidnapped and then arrested, for God's sake. Ted Evans, her former neighbour, had been convicted of a double murder. They deserved some respite.

"Go back inside," she called as she passed the people emerging from their houses. "Stay indoors." They nodded at her, some muttering questions. But no one retreated. Instead, they followed.

She thundered to a halt as she reached the scrubby bushes that flanked this edge of the village. The road ended abruptly, giving way to potholed gravel and weeds as if someone had been interrupted while building it. Not unlikely, given that this place was still under construction when the flood hit six-and-a-half years ago.

Ahead of her, on the horizon, the glow was ballooning, mushrooming into the sky. She took a deep breath then coughed noisily.

Colin Barker was here already, the secretary to the village council.

"Jess." He stepped out from a crowd of villagers peppering him with questions. "There's been an explosion."

She pushed her ginger hair out of her eyes. The night was cold and still, the moon bright over the sea to her right.

"Where?" she asked.

"Not sure. Somewhere in Filey."

She swallowed. "Shit."

"At least it wasn't here."

She turned to look at him. "People could be hurt, Colin."

"Outsiders, Jess."

"Still." She felt a tremor run down her spine.

"Do you think they'd care if something like that happened to us?"

She said nothing. More villagers arrived, jostling her in their haste to get a good view.

"Exactly," Colin said. "They don't give a monkeys about us, so why should we about them?"

"Two wrongs don't make a right, Colin."

"Is that what you said when those men tried to kill Ruth?"

She glared at him. "Don't."

"I know what happened, Jess. I know why they came for her."

She tightened her jaw. Of course he knew. Despite her promise to her brother Ben to keep it secret.

"This isn't the time, Colin."

He grunted. She resisted the temptation to remind him that he wasn't exactly the innocent in all this. Not after he'd locked that boy from Filey up in the boat house – oh yes, she knew things too. Instead, she turned away from him and pushed through the crowd of villagers. Where were Ben and Ruth? Ben, surely, would be somewhere nearby, wanting to get involved.

"Ben." He was with Clyde and Sanjeev, muttering at the edge of the crowd. "Where's Ruth?"

"With the boys. What are you planning to do about this?"

"About what?"

He gestured towards the steadily brightening glow along the coast. "That."

"I can't exactly put it out, can I?"

"It'll be trouble."

"I don't see how. It's miles away."

He shook his head. She glanced at Sanjeev, Ben's best friend. He threw her a tight smile.

"It'll make its way here, sis," said Ben. "You can count on that."

She shook her head. Where was Zack? Had he continued with his little speech, his attempt to justify to his parents why he wanted to marry a woman who was not only the village steward, but nine years older than him?

"Let's get everyone inside," she said. "It might not be safe."

Ben shrugged, his eyes dark, and turned towards the crowd. With Sanjeev and Clyde's help, they started corralling everyone back to their homes.

2

SARAH SAT DOWN NEXT to Martin, feeling the sofa dip beneath her weight.

"How was your mum?" he asked.

She smiled and blew on her mint tea. "Good." She sipped. "Very good."

"New job on the council suiting her?"

"I never knew she had it in her."

"Good for her."

She eyed him. "This is the woman who called you a bad influence. Who wanted you out of the village."

He shrugged. "She had her reasons. She changed her mind, remember."

"She changed her mind about a lot of things."

He said nothing. She drank the rest of her tea, thinking about her father. His trial had been and gone, with no one from the village attending. Now he was in a jail somewhere near Leeds. He hadn't received any visitors.

The sofa shook and she brought her hand up to steady her mug. "Hey, careful."

"I was about to say the same to you."

"That wasn't you?"

"No." He stood and crossed to the window. "Didn't you hear it?"

She placed her mug on the table and slipped in between him and the window, feeling him wind his arm around her waist. Outside was the usual darkness at this time of night. By day this self-sufficient refugee community was alive with activity: people tending the land, making food, keeping the place spick and span. But at night, they retreated. Electricity was permitted for the first hour of darkness, during which time many people sought comfort in the warmth of the JP, the community pub.

But now it was dark, and still.

"Oh my God." She felt her legs tremble. From outside there was a crashing sound, followed by something like the roar of a river.

Except there were no rivers near here.

She grabbed Martin's hand. "Come on."

She pulled him towards the door of the flat and they stumbled downstairs in the darkness. Candles weren't safe in the stairway and oil lighting was too dirty, so all they had to guide them was the dim light from beyond the windows. Tonight the sky was tinged with an unfamiliar shade of orange, reminding her of street lamps in the years before the floods.

She pushed open the outer door to the sound of voices passing.

"That way, look!"

"What was it?"

She peered out, failing to recognise faces. Two men ran past, shouting to each other. She shrank back, scared. They disappeared into the shadows, in the direction of the light. The sky above the rooftops glowed. It trembled and swayed, seeming to be alive with light. She stared at it, open-mouthed.

More people passed; a woman with a teenage boy and another woman with a man not far behind.

Sarah stepped forward, full of questions.

"Stay here." Martin's breath was hot on her neck. "Please."

She turned. "They won't bite."

"I know. But… well, we don't know what's happened. I don't want to be blamed."

"Why would they blame you?" She tried to push the irritation out of her voice. "That's ridiculous."

"I'm not popular. You know it. Please, Sarah. Just stay here. Just for now. Till we know what's going on."

"Skulking in the shadows will just make you look suspicious."

"Please."

His face was close to hers now, his eyes bright against his skin. He was staring into her face, his eyes flitting between hers.

She sighed. "Alright. Of course."

She should be more sympathetic, she knew. But six months ago she would have done what he'd said; or rather what her father had said. She would have stayed indoors, hiding away, never told what was going on outside.

Things had changed.

Martin shouldn't be so timid. He hadn't been like this before, when he had saved her from Robert Cope. When he had helped her understand the truth about her father.

She grasped his hand. It couldn't be easy, being the village's newest and most unwelcome resident. "Sorry. We can stay here. Of course."

She was interrupted by a deep-throated rumbling

coming from somewhere behind them. She turned, realising that she was holding her breath.

"What's happened?" she whispered. She turned to Martin, a finger on his lips. "Two seconds. I'll be back. I promise."

She stepped forwards. A man and a woman were passing, almost running but not quite daring to in the darkness. She recognised them from the school; they had a boy. Craig. Blond, cheeky. Clever.

"Hello," she said, her voice unsure.

The woman stopped. "Sarah. Did you hear?"

She nodded. "What is it?"

"An explosion, over towards Filey. That's what Pam said."

"Pam?"

Pam was the stern woman who presided over the village store. Sarah hated having to go to her to collect her rations every day, knowing that the woman was judging her, gossiping about her to the next person she encountered.

"She was walking home from the JP when it happened," Craig's mum said. "Half the council is out."

The woman peered round Sarah, no doubt looking for Martin. Sarah didn't enlighten her. She felt her breathing return to normal.

Was it bad, that she was relieved it hadn't been something closer? Something Martin might be blamed for? No one in Filey would care about him. At least, not anymore.

"Thanks." She turned back to the flat.

"You're not coming to see?"

"No." Why did people have to be so mawkish? "Thanks."

"Right. Goodnight."

The woman's words were followed by another rumble

from the north. Sarah felt her heart hammering at her rib cage. She needed to get back to Martin. She needed to check on her mother.

"Have you seen Dawn?" Her voice was little more than a croak.

But the woman was gone, loping towards the glow over the rooftops, her hand gripping her companion's sleeve.

Should Sarah go looking for her mother? She was a grown woman after all, one who had suffered through much worse than a distant explosion.

"Sarah!" Martin hissed from the shadows. He'd ventured away from the wall of the building that contained his flat, and was in the shadow of an oak tree. Sarah looked at it, wondering what this land had been used for when it had been planted.

"Sorry." She slipped back to him. The April night was cold and crisp; it wasn't raining, for once.

"What happened?" he asked.

"An explosion, near Filey," she said. "Apparently."

"Blimey."

She drew closer to him, aware of his breathing in the quiet.

"What kind of explosion?" he asked.

She shrugged. "Dunno."

"Maybe I should help."

She put a hand on his arm. "A minute ago you wanted to hide away. Besides, what can you do? It's miles away."

"I don't know. But surely something."

"Please, Martin. Let's go back inside. I don't like it with everyone running around like this."

She patted his arm and he turned back towards the flat.

"Ow!"

"Shit, mate. Sorry."

Martin stumbled, hit by a man who'd appeared out of the darkness, running the same way the first couple had been.

"It's OK." Martin stood up.

"Oh. It's you."

Sarah stepped between Martin and the man. He was young, about six feet tall. He smelt of wood smoke.

"This anything to do with you?" the man asked.

"No," said Sarah. "It's in Filey. Go to the northern edge of the village and you'll be able to see it."

The man stepped towards Sarah as if about to reach for her. She shrank back.

"You're alright, you are," he said. She felt her skin contract.

"Leave her alone," said Martin.

"It's fine," she said. "I can look after myself."

"Yeah," said the man. "She doesn't need you. None of us do."

Sarah narrowed her eyes. "What's your name?"

"What, going to tell on me to your mum?"

She felt her face flush. "No. You haven't got kids,, though."

"No, I haven't. Sorry, teacher."

"You're one of the lads who goes down to the earthworks, aren't you? With the Golder brothers."

"What's it to you?"

"Next time you want to lay into Martin, have a chat with Sam Golder first. Not everything is what you think."

The man snorted, gave Martin a gentle push on the chest, and ran off.

Sarah watched him melt into the night, her breathing shallow.

"You didn't have to do that," Martin said.

She gulped down the lump in her throat. "I don't know where it came from."

Martin put his arm round her and she leaned into him. "Don't put yourself at risk for me. Please."

She said nothing. Martin was the best thing that had ever happened to her, despite everything he'd done. He was worth the risk.

3

Ruth wiped her hand on the tea towel she'd stuffed into the waistband of her jeans, wondering when Sean and Ollie would become less messy at breakfast. She only had twenty minutes before she needed to be at the pharmacy; Sheila Barker had asked to see her at ten. She hoped it wasn't anything serious and wondered if Sheila had told her husband. At least Ruth could be discreet.

She wondered if it was Sheila at the door. Ruth wouldn't blame her; no one enjoyed passing through the village shop into the pharmacy, making it known that they were seeking medical help.

Ruth took a deep breath, pinned on her best smile, and pulled the door open. Outside it was grey, clouds scudding over the houses and a light drizzle beginning to form. In front of her stood a broad, dark-skinned woman she didn't recognise and a thin, pale man who looked as if he'd rather be anywhere but standing on her doorstep on a wet April morning.

Behind them was a police car. Two people sat inside it. Ruth felt her heart lurch.

She looked back into the house – *Ben, where are you?* – and put a hand on her chest. Her body felt light, as if she might float away.

She stared at the woman, waiting for her to speak. A new detective?

The woman smiled at her. *That won't help*, thought Ruth. She turned back to the house again, and croaked out Ben's name. Where was he?

"Ruth! Ruth, what's going on?" Ben rounded the corner of the house; he'd been out. Why hadn't he told her? She felt herself dip with relief.

"I don't know."

The woman turned. She had long dark hair tied back in a bun, the hair pulled so tight it made her look as if she'd had a face lift. The man next to her continued to stare at Ruth. She looked down at her hands; she hadn't realised she was wringing the tea towel.

"What's going on? What are you doing here? My wife was exonerated. You've got no business—"

The woman raised a hand. "We aren't here for your wife. We're looking for Jess Dyer. She isn't at her house. Am I right in thinking you're her brother?"

"What d'you want Jess for?"

"Village business. She is still the steward, yes?"

Ruth watched Ben's face drop a little as he stepped towards the woman. He still couldn't hide his disappointment at Jess taking over his old job.

"She lives two doors along now. In the house where the Evans family used to live. I imagine you know who they are."

Ted Evans was known to the authorities, and he would certainly be known to the two shadows inside that police car. After Ruth had been released without charge, the police persuaded that she'd killed Robert Cope in self-defence, it had been Ted's turn to be taken away. Only in his case, it had been more permanent.

But why were they here now? Had they changed their minds? Was she about to be taken away from her boys again?

She put a hand on the doorframe.

"Ruth, get inside," Ben muttered. "You look pale."

She let out a shaky breath and took one last look at the police car. She recognised the woman in the driver's seat. PC Cregg. A plump young thing, she'd been nervous to find herself in the same room as a suspected murderer. They didn't get much of that round here.

She let herself almost fall back into the house and landed on a kitchen chair, feeling it shift beneath her as it took her weight. She could still hear Ben talking to them.

"She won't be in," he said. "At this time of day she's making her rounds."

"Rounds?"

"She likes to go round the village, visiting anyone who needs help with anything. Checking up on people."

"How long will she be?"

A pause. "No idea."

"In that case, can we come in, please? You may be able to help us."

Ruth stood up. She stared at Ben's back in the doorway. Her heart was racing and her skin felt damp. What did they want?

It had been six months since her arrest. Six months since Robert Cope and his men snatched her from her own home and imprisoned her in that godforsaken farmhouse with Sarah Evans, Roisin Murray and Sally Angus. Since Martin had helped Sarah escape, and come back to the village with Jess's blessing. Since Ruth had stood over Robert, watching blood splutter from his lips as she ground her heel into his chest, the pressure killing him as surely as the knife wound Martin had already inflicted.

He had imprisoned her in his bedroom. He'd touched her, in ways that made her want to climb out of her own skin. He'd intended to rape her.

He deserved everything he got.

She slipped across the kitchen behind Ben. On the staircase, she looped an arm around the bannister, trying not to remember the stairs she had thrown herself up when she had been fleeing Robert. She made her way to the top and stared at her bedroom door, and at the door to her boys' room. They were safely out at school, getting what passed for an education under the tutelage of Sheila Barker and her new apprentice Sarah Evans. She muttered thanks that they weren't at home to witness this.

She heard the rustling of coats and the creak of feet on the wooden floor as the two strangers followed Ben into the house. They stopped in the hallway. Ben didn't invite them in any further.

"Why are you here?" Ben glanced towards the kitchen, frowned, then looked up the stairs. Ruth pulled back, anxious not to be seen.

"We need some help," the woman said.

"What kind of help?"

"Did you see the explosion last night?"

Ruth clenched her fists. What did that have to do with them? Surely they weren't blaming the villagers?

"Yes." Ben's voice was clipped.

"It happened in a residential area."

"Oh." A pause. "I'm sorry to hear that."

"Mr Dyer, please can we come in further? The rain is getting in, your floor—"

"We'll be fine here. You already said it's Jess you want to see, not me."

"You're a member of the governing group in this community, though?"

"The village council. Yes."

"Good. We need your help."

"You already said that."

"Yes."

"What kind of help?"

"The explosion made twenty-six families homeless. With the damage to our infrastructure after the floods of 2017, well, we still haven't recovered."

"Who's we?"

"Filey. The town."

"Oh."

"Look, Mr Dyer. The fact is that you have these large properties here, in your village. They were allocated to you when you arrived here after the floods."

"Yes."

"My colleague here was one of the people you dealt with."

"Not me," said Ben. "I didn't come in the first wave."

"Anyway, it was the borough of Scarborough that generously let you people have these houses. We paid the holiday company handsomely for them."

The village had once been a holiday village, a place for wealthy southerners to spend their summers. Such things seemed ridiculous now, with over a million people being made homeless in 2017. Most of them Londoners, like Ruth and Ben.

Ruth shuffled forward to get a better view.

"Are you going to get to the point?" Ben looked as if he was about to push the woman out of the door.

"We need your houses, Mr Dyer. We need to house the people of Filey."

"You what?"

"You heard me."

Ruth put her hand to her lips. She couldn't see Ben's face, but she could hear his reaction in his voice.

Ben said nothing. Ruth heard movement and spotted the woman moving closer to him. She tensed.

Ruth thought back to four years earlier, when the people of Filey had attacked the village. They'd had to post guards; they'd feared for their lives. The boys were tiny babies and Ruth had barricaded herself into the nursery with them. It had been terrifying.

And now they wanted to house them here.

4

JESS'S ROUNDS took longer that morning. Everywhere she went, she was asked about the explosion. What had caused it, where it had been, who had done it. As if she knew. It hadn't taken place in their village, and didn't seem to have involved anyone from here. That was all she was going to let herself worry about.

But as steward she was expected to have answers. She wondered if people had looked at Ben so distrustfully. He'd served three years on the council before he'd actively pursued the steward job, glad-handing people who didn't even have a say in the election, then manoeuvring for the succession after Colin Barker's term had expired. And he'd loved the job. It had given him a purpose she'd never seen before: not when they were growing up, not when they'd been forced from their homes by the floods, and not on the journey here.

He seemed to have forgiven her now, but she couldn't be sure. She would have to content herself with doing her best and hoping it was enough. Would anyone ever forget she was the steward who had insisted on mounting a rescue mission at sea, who had brought those men to their village, who had offered to give them shelter? She'd been repaid by having her sister-in-law and three other women taken from them. They'd got them back, but at what price?

She would take half an hour's break, have a quick cup of tea and rest her feet. She'd gaze out of the window of her new house towards the sea, and try not to think about the things that had happened when its previous occupiers had been there. She knew some of it; she'd found Dawn slumped over the sofa she now forced herself to sit on every evening, and had helped Dawn's daughter Sarah carry her upstairs. Ted had been a controlling man, and a brutal one. But now he was gone, and Dawn had chosen to take a smaller house in the centre of the village.

The house felt more like home when Zack was with her. Jess felt a flood of warmth trickle through her at the thought of him. She was steward; she'd have to ask another member of the council to marry them. Colin would make the perfect candidate; after all, he'd done it for Ruth and Ben during his term as steward.

She left the Meadows behind and shook the mud off her boots, heading for the houses where she and most of the other council members lived. She stared at the ground as she walked, measuring her steps.

Something was different. A sound, overlaying the familiar repeating tones of a song thrush in the hedge and the waves behind the houses.

She stopped walking. It was a car engine, idling.

She picked up her pace and rounded the corner to her row of houses. A police car stood outside Ben and Ruth's house, with two shadowy figures inside.

She looked to her left, towards the entrance to the village. There were no other cars. Just this one police car.

Where were the others? When they'd come before, there had been at least two each time, sometimes four.

She dropped the bag of potatoes she'd been carrying, a gift from Flo Murray, and started to run. She ran into Ben and Ruth's door at full pelt, hammering on it.

The door flew open and she almost fell through.

"What's going on? Where's Ruth?" She was panting.

"It's alright, sis," said Ben. "It's the council."

"The council?"

"The local council. Not the police."

She flung an arm in the direction of the police car. "So what's that then?"

Ben screwed up his face. "Protection."

"Protection? For who?"

"For them. The council officers." He looked over her shoulder. "Come in. They want to talk to you."

She shook herself out and stepped inside, shucking her muddy boots off. Two strangers were in the living room, a large woman perched on the edge of the sofa and a balding man on the armchair.

She licked her lips and strode in, trying to project confidence.

"I'm Jess Dyer." She held out a hand. "What brings you here?"

The woman stood up. She smiled at Jess and shook her hand. As she did so she cast a wry look in Ben's direction.

"Glad to meet you, Ms Dyer."

"Who are you?"

"My name is Anita Chopra. I'm from Scarborough borough council."

Jess stiffened. Had they heard about Ruth's arrest, and Ted's? Did they know about Martin? She wondered if she had the authority to let him live here. Sarah would be devastated if he was evicted.

"Shall we take a seat, Ms Dyer?"

Jess glanced at Ben, feeling like a girl pulling faces at her brother behind their parents' back again. "No thanks, I'll stand."

"Oh. Of course."

The man, who had been perching on the armchair and staring out to sea, stood up and joined his colleague. Jess looked him up and down; he looked like he needed a good meal.

"Go on, then," Jess said.

"I've been explaining to your brother that we have a bit of a situation in Filey."

"A situation?"

"The explosion last night. It happened in a residential area."

"I'm sorry to hear that."

It was true. However little time Jess had for this woman and her pale sidekick, however little she trusted the authorities who had never lifted a finger to protect them from their tormentors. If people were hurt, it wasn't their fault.

"Is anyone hurt?" she asked.

"Five people have been taken to hospital in York. Plenty more have minor injuries. But it's the housing situation that I have to deal with."

"The housing situation."

"Twenty-six families have been made homeless. We need to bring them here."

"Here?"

"You have plenty of space."

"All our houses are occupied."

"Not fully occupied."

"That's hardly the fault of—"

"I'm told you've moved to the house two doors up, is that true?"

"Yes."

"A three-bedroomed house."

"Yes. We find that if council members live near the centre of the—"

"And you live alone?"

"With my – she hesitated at the word *fiancé* – my boyfriend."

"I'm sure you'd agree, Ms Dyer, that it seems unfair for a woman on her own to be living in such a large house when there are people sleeping on the floor of Filey church hall."

Jess narrowed her eyes. She'd slept on the floors of church halls herself, after the floods had wrecked her London flat. In London there had been sports halls and indoor stadiums, thronged with people. On the long walk north, they'd taken everything they could get. Including church halls.

"I'm sorry, Ms Chopra."

"Miss Chopra."

"Miss Chopra. It's not my decision."

"You're the steward."

Ben stepped towards them. "We run things democratically round here."

Miss Chopra pursed her lips at him. "I'm sure you do. But this is an urgent situation. Surely you can find space for these poor people now, and then decide whether you can offer them sanctuary for a longer period?"

Jess saw Ben stiffen. *Sanctuary*. That had only led to trouble, last time.

"No," said Ben.

Jess clenched her fists. "Ben, please. It's not up to us."

"You know what they'll say. They know what happens when we let outsiders in."

"We've let Martin back in."

"Have you seen the way people look at him?"

She turned to the council officer. "I want to help you, really I do." Was that the truth? "But like my brother says, we run things democratically. We'll have to consult with the other villagers, and let you know."

"There's no time for that."

"I'm sorry?"

The woman returned to the sofa and grabbed a bulging briefcase that was leaning against it. She drew out a file.

"If you don't cooperate, I can go to a judge. We gave you these houses, we can take them away again."

Jess closed her gaping mouth. "You wouldn't."

"I don't want to. Really, I don't. But we have people who need a place to live."

"Please," said Jess. "Give us twenty-four hours. I'm sure I can persuade—"

"No," said Ben. "She can't. No one is persuading anyone of anything. I always knew we couldn't trust you lot. Get out."

"Ben—"

Ben put a hand on Jess's forearm and squeezed. It hurt. She tried to shake him off but he wouldn't budge.

"Ben. Please. Let's be reasonable."

"No, Jess. This is my house – for now – and these people aren't welcome." He turned to their visitors. "Leave, please. Now."

5

THE HOUSE SHUDDERED as Ben slammed the door behind the two council officers. Ruth leaned into the wall, shrinking away from the voices that came up the stairs.

Ben stood with his back against the door, his chest rising and falling. Ruth watched him, wondering if he'd thought to consider the things that had happened to her.

"That was stupid." Jess's voice was jagged with anger.

"They can't kick us out."

"Can't they?"

"We've been here for six years. That counts for something."

Jess moved to the bottom of the stairs. Ruth considered going down to them, joining in. She shivered and decided to stay put.

"It counts for nothing, Ben," Jess said. "Have you read the housing agreement?"

"I was the steward. Of course I have."

"Then you'll know the eviction terms."

"They have to give notice."

"Unless…"

Ben glanced up the stairs. Ruth held her breath, confident he couldn't see her in the dark.

"Unless what?" he snapped.

"There are clauses about what happens if anyone from the village is convicted of a crime."

"Ted's gone. That can't affect us surely."

"Ben, don't be naive. They'll use it against us. If we don't cooperate, they could sling more of us out."

"They're already slinging us out!"

"They aren't. They just want to make better use of the housing stock available to them."

"Housing stock? Is that all this is to you? Has moving out of the house where Mum died made you so callous?"

"That was low, Ben," Jess said.

"This isn't housing stock. It's our homes. It's where Ruth and I are raising our boys."

Jess's shoulders slumped. "I know that. But I don't think we can just turn them down. It might—"

"Do you want them here? More outsiders, threatening us?"

"No, of course I—"

Ruth could feel her skin frosting over. She felt sick. She closed her eyes and leaned back. *Go, Jess, please. Leave us alone.*

"Don't you remember those boys," Ben said. "On your first day as steward? The way they threatened those children? Told us to *fuck off skum*?"

"They were just kids."

"They get it from their parents, and you know it."

Jess said nothing. Ruth heard her move into the kitchen. The tap ran and then there was a dull thud as the pipe cleared.

"Don't walk away from me when we're having a conversation." Ben's voice was fainter now; he'd followed his sister into the kitchen. Ruth slipped down a couple of steps, her guilt at eavesdropping cloaked by anxiety.

Outsiders. Last time they'd allowed outsiders into this village, it had been Robert and his men.

"There's no point," said Jess. "It'll have to go to a vote, anyway. This is too big for you and me to decide."

"OK. I'm sure the council will see it my way."

"Not the council. The whole village. This is too big. People could be made homeless."

"That's why we have to say no."

"I'm not so sure, Ben. If we refuse, it could be a lot worse."

6

BEN SAT at the kitchen table, his fingers clenched around a mug. Ruth watched the rise and fall of his shoulders. His muscles were slack, his body slumped.

She picked her way across the room and placed a hand on his back. He flinched.

"Ben."

He turned, eyes dark. He looked tired. She felt her heart fill with concern and love.

"Are you alright, love?" she asked.

"Did you hear all that?"

She nodded.

"They don't get it. *She* doesn't get it."

Ruth dragged a chair out from under the table and positioned it to face him. He grabbed her hand and brought it onto his lap. He stroked the skin on the back of her knuckles. It tickled; she resisted the urge to pull it away.

"Have they gone?" she asked.

"Who? Those council officials? Bloody do-gooders."

She shook her head, feeling her chest tighten. "The police."

Ben's head jolted up. "Oh. Ruth, I'm so sorry."

She shrugged. *Don't say it. Don't mention it.*

Ben shifted his chair towards hers and leaned forward to put his arms around her. "I'm sorry, love. I didn't think."

She blinked, glad he couldn't see her face. "It's alright."

He pulled back. "You're so brave."

She bit her lower lip and nodded.

"You thought they'd come to arrest you again."

She frowned. There had been a moment, a fleeting moment, when that had passed through her mind. But no; that wasn't what she was scared of. Not really.

But he didn't need to know that. She pulled his hand up to her face and rubbed it against her chin. The skin on the back of his hand was rough; since losing his job as steward he'd started working on the allotments, and now he had gardener's hands. It felt real, comforting.

"I don't want them here," she said.

"Who? The police? The council?"

"Anyone."

"You mean the people they want to house here."

She swallowed.

"I'll do my best, Ruth." He pulled back, examining her face. "I thought you'd argue against me."

She frowned.

"After you helped Martin with his hypothermia. You let him stay here. Then you treated him, after he came back again. You've always been so generous. So welcoming."

"That was different."

"Was it?"

"Yes." She blinked, trying to push the thickening clouds from her mind. *Don't think about it*, she told herself. *Forget. It's over.*

Except it wasn't. New people were coming here, people

who hated them. People who'd attacked their community in the past, who'd told them to *go home*.

How could they possibly coexist, if those officials got their way? How could they survive?

"You're right," she said. "I used to think this place was a sanctuary. But now I know."

"What do you know?"

"It *is* a sanctuary. But only if we keep it safe. Only if we don't let anyone else in."

7

Jess hated listening to Pam. As the proprietor of the village shop and the keeper of rations and supplies, Pam was party to almost everything that went on in the village. Every family in the place came through her doors most days to pick up their rations. She never missed an opportunity to pry.

But Pam was powerful. She resented the council members and their privilege. The well-placed houses, the power over people's lives.

Jess often wondered if Pam would exercise her own power differently, the power she wielded over food, if they hadn't put checks and balances in place right from the start. Pam distributed rations. But Toni was in charge of getting food to the shop from the allotments, the smoke house and the bakery. She and Pam went over everyone's supply each day, checking and double checking that everything was shared fairly. The two women hated each other. But maybe that helped. Maybe that kept them from working together to abuse the system.

Jess had bumped into Toni on the way here; her friend

had been taciturn and gloomy. When Jess had asked her what was wrong, Toni had shaken her head; nothing for Jess to worry about. She'd muttered Roisin's name but said no more. Jess knew she shouldn't pry, but Toni had been distant lately. Jess was worried about her.

She needed to be more trusting, to worry less. But it was hard, after the way that woman had talked to her. Anita Chopra. Scarborough council, the body that had left them here to rot.

They'd adapted, partly thanks to this system that Pam and Toni presided over. But that didn't make it any easier.

Jess turned at the sound of the bell over the shop door, relieved to have some respite from Pam's voice. She knew she should listen when Pam gossiped. There might be information there that she needed to know. But there would also be information she most definitely did not need to know. It wasn't the steward's place to delve into people's intimate lives. So whenever she had to come to the shop, she drowned it out with her own thoughts. Mainly of Zack.

The door opened and Ruth entered. She flinched as she spotted Jess, then put on a smile.

"Hello."

"Hi, Ruth. I've been looking for you."

"Oh."

Ruth's eyes wore dark circles and her cheeks were puffy. She looked as if she hadn't washed her normally thick, dark hair for days.

"Has Ben told you about the council officials who came round earlier?" Jess asked.

Jess sensed Pam prick up her ears.

"Come into the pharmacy," Ruth said.

Jess gave Pam a brusque nod and followed Ruth through.

Inside, the pharmacy was still and quiet. Boxes were

arranged neatly on the shelves and the table in the centre of the space had been wiped clean. The walls held two posters on first aid and the floor was clean and free of footprints. There was a faint smell of vinegar and elderflower.

Ruth waited for Jess to close the door. "Go on."

"Sorry about this. I know you don't need me asking more of you."

"I'm not busy right now."

"I know, but—"

"Please. Tell me what it is you need."

"OK. It's the explosion last night. They need medical help."

"Oh."

"They asked if we had a doctor."

"We don't."

"Ruth. You may not be trained as a doctor, but you're the closest we've got."

"They don't need me. I'm just a veterinary nurse."

"I think the last few years have qualified you for more than that. You're an expert at practising medicine without traditional supplies."

"I wouldn't say that."

"You are, Ruth. I can smell it, in this room. I know you struggle to source medicines. I know you go down to the beach, and into the fields. You use herbs, and minerals from the rocks. It's impressive."

Ruth shrugged. "It's not medicine."

"Try telling that to all the people you've treated."

"It's not. They'll have hospitals, and real doctors. They don't need me."

"They need all the help they can get. There are a lot of people with injuries. They need to stem infection, and they have as much problem accessing antibiotics as we do."

"Antibiotics that still work, you mean."

"Yes." Jess could feel her pulse rising. This wasn't like Ruth. "Can you help?"

"I think I'd do more harm than good."

Jess's shoulders slumped. Ruth turned her back, shifting boxes from one shelf to another; something Jess was sure she didn't need to do. She'd probably move them back again after Jess had gone.

"Please, Ruth."

Ruth kept her back to her sister-in-law. "I don't think it would be wise."

There was a knock on the door behind Jess. Ruth pushed past her and opened it. "Sheila. Good to see you." She turned to Jess. "I need you to leave, please."

8

Sarah listened to the voices outside; friends and neighbours, all heading for the same place. A child shouted something and a woman laughed. She went to the window and looked out. A family had stopped below, waiting for a little boy to catch up. Ezra Clarke: she knew him from the school. He shrieked something unintelligible and then ran to catch up. His mum ruffled his hair and continued walking.

"I think we should go," she said, still looking out.

"I'd rather not," Martin replied.

"Mum will be wondering why we're not there."

"She'll be too busy."

Sarah turned and sat on the sofa next to Martin. "You're part of this community now. You'll be with me. You shouldn't let them scare you."

He shifted away from her. "They don't scare me."

"Why won't you go, then?"

He took a deep breath. "I didn't tell you what happened this morning."

"This morning?"

"Yes."

"Go on."

She kept her smile on, not wanting to make him any more worried. This wouldn't be the first time Martin had been met with hostility.

"It was Mark Palfrey," he said.

"Who?"

"Sally Angus's boyfriend."

"Oh. How is Sally?" Sarah felt a pang of guilt; she hadn't spoken to Sally more than twice since they'd been kidnapped.

Martin eyed her. "I don't bloody know, do I? Sorry, that sounded callous. Mark's not exactly going to tell me, though. He made it pretty clear how he feels about me."

"What did he say?"

Martin pursed his lips and let out a whistling breath. "I'd rather not say."

"You were the one who wanted to tell me about it."

A pause. "Alright. He told me to go back to the farm. Said no one was safe with me here."

She put a hand on his arm. "I'm sorry."

"It's no more than I deserve."

"It's not. Robert made you do what you did. And you rescued me, didn't you? You stopped Robert."

He was chewing his lower lip, saying nothing. She squeezed his arm, feeling the muscle tense beneath her hand.

"That's not you. I know what you're like. My mum does. Jess, and the other council members."

He wiped his face. "They're not what counts though."

She felt suddenly irritated. She let go of his arm and stood up.

"Come on. You have to face them. You have to show them you're not what they think you are."

"Not tonight."

"Why on earth not?"

"Because this meeting is about bringing outsiders into the village. People like me, as far as they're concerned."

"People who've been made homeless. Just like we were. Families."

"They won't see it like that."

She was at the door now. Her coat was on the hook next to it. Should she go without him, support her mother?

"Martin, there's only one way to know for sure, and that's to come."

He shook his head.

"Please." Her hand was on her coat. She tightened her grip around it but didn't pull it down.

He looked up. "You go. Be with your mum."

"I don't want to go without you."

"I don't mind." But he did; she could tell from the way he was holding himself, turned away from her and stiff.

She let go of her coat and returned to the sofa. "I'm staying with you."

"No. You go. You don't need me holding you back."

"Come with me."

He crossed to the door and yanked her coat down from its place. He threw it into her lap. She grabbed it, irritated.

She stood up. "Alright then."

She headed for the door and clattered down the stairs without looking back. There were just a few stragglers passing now, people who in all likelihood would be late for the meeting, like her. She hoped they hadn't shut the doors.

She ran to the village hall, grabbing her skirt in her fist so as not to trip. A fine drizzle was beginning to fall and dusk was descending. Somewhere in a hedge, a dunnock was calling its mate. A few windows glowed into the night.

Not many though; no one here would leave electricity on while they were at a meeting.

Sanjeev was just closing the doors. She squeezed through, shooting him an out-of-breath smile. He looked past her, clearly wondering where Martin was.

Inside, the room was full. People stood behind the rows of chairs, jostling to get a view. She slipped between two men and found a space at the side of the room from where she could see her mother. Dawn was sitting in the front row, right at the end. She had a notebook in her lap and looked pleased with herself, as if proud of her role here.

Until six months ago, Dawn had been a ghost to the other villagers, someone who they rarely glimpsed and hardly ever emerged out of her own front door. But now Sarah's father was in prison, and Dawn was an active member of the village council.

Someone jostled Sarah and she looked round. It was Zack Golder, Sam's twin. She looked past him to see if her friend was with him.

"Hi Sarah."

"Hi Zack."

"No Martin?"

"No. Where's Sam?"

"Can you close the doors please Sanjeev?" called Jess Dyer from the front. "I'm worried it's getting overcrowded."

The doors closed with a thud and the room grew quiet. The woman in front of Sarah shifted a little, giving her a better view. Jess was perched on a table at the front, facing the rest of the village. Colin sat next to her, looking uncomfortable; he'd rather use a chair, she imagined. But there wasn't space.

Sarah liked Jess. She trusted her to come up with the right solution to this. Their community was wary of outsiders, and rightly so. But surely the people of Filey deserved the same refuge they'd enjoyed here?

"Right," said Jess. "I imagine you all know why we're here. I know how quickly news spreads."

There was muttering around Sarah. Colin put his hand up for quiet.

"You saw the explosion in Filey the night before last. It was a gas main, in a residential area. Two people were killed, and more injured. Twenty-six households have been made homeless.

The muttering rose.

"You all know how this village started. Every one of you was displaced by the floods. You had nowhere to go, and this place took you in. I think we should continue that tradition. What does it make us, if we turn them away?"

"Would they do the same for us?" A voice at the front.

"I don't think that's the issue," replied Jess.

"They hate us!" Another voice, closer to Sarah.

"Why should we be as bad as them?" That was Zack, standing right next to her. Sarah threw him a grin. Maybe if they took in a few more newcomers, Martin would be less noticeable.

"Where would they live, if we said yes?" Ben Dyer was standing up at the front, flanked by his two boys, their blonde hair bright in the dimness of the room. Beyond one of them – Sean, Sarah thought – Ruth sat, her back straight.

Jess pursed her lips. "We would need to make more efficient use of the space we have."

"Kick people out of their homes, you mean," he replied.

"No. There is the option of asking for volunteers."

"Volunteers? How many people do you think will

voluntarily leave their home, after everything we've all suffered?"

Jess dipped her head as Colin whispered something in her ear. She frowned.

"I won't!" cried a woman behind Sarah.

"Nor me!"

"Me neither!"

"*You* should!"

Jess's head whipped up.

The same voice came again. "The steward lives in a big house, on her own. She should give that up. Maybe for some of us. We've got six people in our house. It's hardly fair."

Jess nodded. "You're right." She looked at Zack. He nodded. "I can give up my house," she continued.

Ben was on his feet again. He turned to the crowd. "Can't you hear what they're asking us to do?" he cried. "It's ridiculous!"

9

JESS STARED at her brother's back, her breath shallow.

Not again, Ben.

She knew he didn't agree with her on this. She knew he was scared of outsiders, not that he would ever admit it. But to override her like this, in a village meeting…

"Ben, please..." she began. He ignored her.

"I want to remind you all of something," he said. His voice was low and flat. He didn't make the effort to raise it, and in return, the room descended into a hush so they could hear him.

"Do you remember the twenty-eighth of October last year?"

He paused, his head moving from left to right as he scanned the room. There were a few mutters of *yes*. Flo Murray, three rows back, let out a gasp.

"Of course you do," he continued. "That was the day I did the most stupid, regrettable thing I've ever done."

Silence. At the time, he had blamed her. After all, she had convinced him to answer that distress call at sea.

"I allowed those men to come to our village. Robert Cope and his friends."

He looked round at Jess. She stiffened. He turned back.

"Thank you, Ben," she began. "Can we please get on with the—"

He raised a hand to dismiss her. She gritted her teeth. How dare he?

"I take responsibility. I let Jess persuade me to send villagers out to answer that rescue call. I sat in a council meeting where we agreed to give the men shelter. I took one of them into my home."

Jess noticed Sarah Evans, standing just behind Zack. She wondered if Martin was with her. Sarah's face was a deep shade of pink.

"And I will regret those decisions till the day I die."

You're not telling them the rest of it, she thought. *How it was revenge for what you did to him when you were young that brought Robert here. How you helped him kill a man.* She wondered how many people here knew. Sarah, and Zack. Ben and Ruth, of course. Sanjeev. Dawn? Would Sarah have told her? She'd probably not even heard them talk about it, at the time.

"So I'm urging you now," Ben continued. "Don't make that mistake twice."

"This is different," she interrupted. "Families. Who have been genuinely displaced. They aren't faking it, like the men were with their boat."

He kept his back to her. "That doesn't matter. They're outsiders. Outsiders are trouble. They hate us. Don't you all remember four years ago, when we had to place guards on the village perimeter? They would have murdered us in our sleep. They've always resented us being here, and they always will. And if we let them come here, they'll find a way to turf us out. That is, if we're still alive to *be* turfed out."

In the front row, Ruth leaned over Sean and put her hands over his ears. Ollie, his twin brother, let out a sob.

Ben, thought Jess. *Can't you see what you're doing? Fomenting panic. Exaggerating the risk.*

She stood up. The crowd was pushing in on her now, and she could barely see past the front row. Sarah Evans had disappeared into the melee. She hoped the girl was all right.

This will never work, she thought. She climbed onto the table she'd been perching on. Below her, voices were raised. Someone screamed. Jess could smell fear: the acrid smell of sweat.

"Everyone!" she shouted. "Calm down, please. Ben is exaggerating."

Ben turned to her. "Am I? Am I really?"

She glared at him. "Yes. Sit down, please."

"You sit down."

This was starting to remind her of her childhood, the way the two of them had bickered at every opportunity, both of them insisting to Sonia that the other was in the wrong.

"Alright," she breathed. She squatted on the table then lowered herself so she was sitting on it. Ben returned to his seat. Ruth was staring ahead, her jaw set. She'd gathered both boys onto her lap and was stroking their hair, not noticing their protests as her fingers caught in the tangles.

Ben shifted his chair to face sideways, with a view of both Jess and the room. He was enjoying this. Ruth clearly wasn't.

Colin cleared his throat. In all the commotion, Jess had forgotten that he was still standing beside her.

"I think we can have a reasonable conversation about

this," he said. He had a pencil in his hand and he was twisting the tip into the flesh of his thumb. He hated outsiders too. When two youths had invaded their village six months ago, he'd locked one of them in the boathouse.

"Thanks, Colin," she said. She scanned the crowd. Surely there was someone who would support her. Her eye caught on Zack. No, too obvious. She didn't want this looking like a setup.

Sanjeev was right at the back. He tended to be reasonable. But he was loyal to Ben.

Then she remembered. The woman who'd let Martin come to the village to be with her daughter. She turned towards Dawn, sitting with her hands in her lap at the end of the front row.

"I'd like other members of the council to have their say," Jess said. "Dawn, what do you think?"

Dawn reddened. "Well..."

"Remember we're talking about families here. Children, who've lost their homes just like we did." Did she dare try a Bible reference? They appealed to Dawn. But then, she'd probably get whatever quote she tried to drag up from her memory wrong.

"Dawn?"

Dawn nodded. "The poor little mites. They're innocents."

"Exactly."

"But what about their parents? Who's to say they aren't the people who tried to have us evicted four years go?"

Jess sighed. Ben stood up. He turned away from her again.

"Dawn's right," he said. "If we let those people live among us, what will they do to us? It's all very well thinking of their children. What about *ours*?"

He grabbed Sean's hand. Sean grinned and threw his

arms around his dad. Older than Ollie by a full hour and always the confident one, he looked up to his dad. Ollie, on the other hand, was Ruth's baby. He shrank towards her, almost as if he were trying to burrow inside her. *Don't single him out too, Ben,* Jess thought.

But Ben knew his boys better than she gave him credit for. He threw Ollie a smile and lifted Sean onto his shoulders. Sean giggled.

"It's up to you all," Ben said. "But I speak as someone who almost lost his family the last time this happened." He paused. Sean grabbed his hair where it was thinning at the top. "Let's not allow it to happen again."

Colin motioned for Ben to sit down. Ben swung Sean down onto his lap and gave the boy a noisy kiss. He looked at Ruth but she was staring ahead, her eyes bright.

Jess watched her. Was Ruth OK? Had something happened between them?

"Right," said Colin. "It's only fair to ask if anyone else wants to speak."

Jess cleared her throat.

"Anyone else," Colin muttered in her ear. She clenched her fist.

No one spoke. A few people looked at Dawn, then at Sanjeev. But neither of them had anything to say.

"Right," said Colin. "In that case, let's put it to a vote."

10

"We won't let it happen, sis."

Jess stared at him. *How can you be so naive, after everything that's happened to you?*

"They're going to get a court order," she told him.

Ben snorted. "That'll go nowhere. By the time they've done that, those people will have all found somewhere else to live."

"Don't be so sure."

He shook his head. "You'll see. We made the right decision."

"We didn't make any decision. You talked the village into making the wrong choice."

"Hi, Jess. What's happening?"

Jess looked past Ben to see Ruth coming down the stairs.

"Why didn't you say anything, Ruth?"

Ruth shook her head but said nothing.

"You're the most generous, decent, tolerant person I know. You treated Martin after he kidnapped you. Even Ted. You—"

"I don't want to get involved, Jess."

Jess felt herself deflate. "Why not?"

"You don't understand."

Jess bit her lip. Her sister-in-law was right. How could Jess understand everything Ruth had suffered?

Ruth hadn't spoken about her arrest since she'd returned from Filey. As far as Jess was aware, she hadn't even told Ben what she'd been through.

There was a wail from upstairs. Ruth tensed. "See you, Jess. We need to be alone, right now."

Jess blanched; the rejection was like a slap in the face. The three of them had walked here together from London. They'd cared for Sonia together. They'd shared everything. And now Sonia was dead, and Jess wasn't wanted.

She eyed Ben. He said nothing.

"I hope you know what you've done," she said.

He shrugged and opened the front door. She retreated through it, feeling empty. He'd told her he was happy for her to be steward now, that he wanted to focus on his family. But his ambition, his need to be in control, was strong.

Outside, the night was cold. Not many people ventured outdoors in the dark, there being no street lamps. And the winds could blow you right off the cliff edge if you weren't careful. But Jess only lived two doors away and besides, her eyes had grown accustomed; she could see by the starlight. This would be what life had been like centuries ago, before electric light was a glint in anyone's eye.

She dragged her feet towards her own house, reluctant to go inside. Zack was with his parents tonight – they hadn't reacted well to the news of the engagement, and he was trying to placate them – and her house would be dark and cold. She'd got used to being on her own, even in her last house with its memories of Sonia. But here, in this gloomy house with the sea never far away and the wind whistling in the eaves, she experienced an emptiness that felt like it could never be filled.

It didn't help that she could only imagine things that had happened in the house before her tenancy. She'd been a witness once, picking Dawn up from her position slumped against the sofa and helping Sarah to carry her upstairs. How many times had Ted beaten Dawn, behind the door she now called hers? How many times had he terrorised Sarah?

She opened the door and peered inside, half expecting Ted to rush out at her. She shivered and fumbled in the chest of drawers behind the door, searching for a match. She lit the candle she kept on the chest and instantly regretted it. It threw long, flickering shadows across the wall and up the stairs, the kind of shadows that were more at home in ghost stories.

Zack would be back in an hour or two. Maybe she'd go for a walk, to occupy herself.

She grabbed her heavy coat – Sonia's coat – from behind the door and shrugged it onto her shoulders. The weight of being steward felt like an animal draped across her back. Anita Chopra had barely disguised her anger this afternoon, when she'd arrived at Jess's door and been told their decision. She'd left in a flurry of hissed accusations and threats. One way or the other, this wasn't going to be good.

The sky was clear enough now for her to make her way down to the beach safely, and the rain had subsided. She knew where she was going; she'd be careful.

She shuffled along the road that led to the cliff path, imagining what this would have been like when it was a holiday village. Windows would have been lit, people would have been lighting barbecues, cars coming and going. And children. This was the kind of place that would have had children everywhere. At least the children who lived here now could enjoy some sort of innocence. They were allowed to roam the village in a way she'd certainly never been allowed to roam the streets around her mother's London house; and they'd never known twenty-four-hour electricity, the internet, or computer games. But then, who knew what the future held for them? Tending the land and maybe working down in Hull on the earthworks, if those still existed? Not the greatest of prospects.

She heard a yell from somewhere behind her and stopped walking. She listened, breathing in the sea air. Was someone calling her?

She turned back towards the village centre, peering into the night.

Then she heard them. Footsteps, coming from behind her, the direction of the beach. Not just one set, but two; three; more.

She span to face the beach. Who was out there?

"Hello?"

No one replied. Instead, the footsteps grew until she couldn't tell how many pairs of feet they belonged to. They sounded chaotic, like the rumble of feet in a running race. She felt her breath catch.

"Hello? Who is it?"

"Go!"

She frowned and shrank back. "Zack?"

There was a time when Zack and his bother Sean might have played a practical joke on her, back then they were young men. But Zack wasn't stupid; he knew how intimidating the night could be.

"Who is it? Stop it, now!"

An individual set of footsteps approached, echoing off

the darkened buildings. Suddenly overcome by dread, she darted into the gap between two houses.

She stared out onto the road. Dark figures rushed past, feet hammering on the tarmac. Torches flamed in people's hands, raised into the night air.

She stumbled. No one here used torches like that. It was a waste of resources.

Should she make herself known, challenge them? Or not?

The shapes passed, flames flickering off the buildings and the blank windows. She wondered if there were people in there, staring out, hearts hammering in their chest like hers was.

When the light had receded she dragged herself out on to the tarmac. A dimly-lit crowd of people moved in the direction of the village centre. None of them spoke.

She thumped on the door of the house closest to her. A man opened the door, wearing a vest and a pair of patched-up tracksuit bottoms.

"Jess?"

"We're under attack," she panted. "We have to defend ourselves."

11

"WHAT WAS THAT?"

Ruth flinched as Ben jumped up from his place on the sofa and ran to the window in the kitchen. They'd been gazing at the stars through the window in silence. The boys had just gone to sleep and she had no energy for anything more than slumping on the sofa and staring out at the sky. She felt so tired, she wished she could close her eyes and find herself in her bed.

"What was what?" she repeated.

"There's people out there."

She drew a breath and pushed herself up from the sofa. Ollie would be awake again at five. She needed her bed.

Ben was leaning over the kitchen sink, peering sideways out of the window. Ruth could make out dim lights, dancing in the gloom.

"What's that?"

"No idea." Ben turned and stared into her face for a moment. His eyes were full of excitement.

"Careful," she said, as he hurried to the front door. "Please, love."

"It's fine." He pulled on his coat and eased the door open. Light came through the gap, illuminating a slice of the wall. She gasped.

"Who is it?"

"I'll find out. You stay here." He looked up the stairs, the boyish anticipation falling from his face. "Look after the boys."

She nodded; as if she needed asking.

He looked back at her as he stood in the open doorway. "Don't open the door."

She frowned; why not? But then she nodded, swallowing her fear. "I love you."

"You too."

He eased the door closed and was gone. She hurried to the kitchen window but there was nothing to see. Vague figures flashed past in the direction of the square, but the view from here was of the side of Colin and Sheila's house, not of the street.

She hurried to the stairs and took them two at a time. At the top was a window; the one window that faced the road. The building had been designed to make the most of the sea views and so it felt isolated from the life of the village sometimes, with its aspect over the cliffs and the crashing waves beyond.

Outside, lights reflected off walls and windows and there was shouting. She heard a crash and a scream. She pulled her fist to her chest and glanced at the door to her sons' room.

She leaned forward to see better. People were running around in the square. A shape was moving over by the JP, the village pub. She couldn't tell if it was one person lying on the ground, or several people wrestling.

She looked down at the front door. If she bolted it, Ben wouldn't be able to get back in. But if she didn't, her boys were at risk.

Don't open the door, he'd said. *Look after the boys*. There was roaring in her ears that was nothing to do with the noise outside. Her palms were sweaty. Her stomach felt like it was being scrubbed from inside.

She descended the stairs slowly, one at a time, and reached out for the bolt. She slid it closed, then checked it. She pushed against the door, then tugged it.

It was firm.

She retreated up the stairs, not taking her eyes off the door. Soon she was by the window again. Something was on fire out there, but she couldn't make out what. It was at the back of the JP: the composting bins?

She flinched as a knock sounded at her door. She shrank back and stared at it.

Another knock. And another. She stared at it, confused. Could it be Ben?

She stumbled down the stairs. She crouched behind the door, wishing there was a spy hole.

"Who is it?" she croaked.

No answer. The knocking had stopped. She put a hand on the bolt, considering opening it. *No*. She had to stay safe. She had to keep them out.

She pulled herself back up the stairs, sliding up like Ollie had when he was too little to walk. At the top, the first door was the one to her sons' room. She had to protect them.

She pulled herself up and turned the knob, slowly. She eased the door open.

The curtains were open and shapes danced on the walls. Dim, red shapes. She stumbled to the window and pulled the curtains closed, not pausing to look out. No one could know her children were in here.

She lowered herself to the floor between their beds, thinking of the night Martin had slept here, in the room beyond the wall that ran next to Sean's bed. She'd been worried about their safety, wary about the newcomers. So she'd slept in here, on the floor. And he'd taken her from here.

She lay down on the floor and stared up at the ceiling. Her mistake that time had been to fall asleep. If she was unconscious, she couldn't protect her children. She would keep her eyes open until the morning came, or until the noise outside stopped. Whichever came first.

12

Jess ran to the village square. The houses were dark, the night chilly. She looked back at the council members' houses; she needed to alert them. Why had she run past them?

"Jess!"

She span to see Toni running towards her.

"Toni! Have you heard?"

"Heard what?"

"We're under attack. People from Filey, I think. Wasn't that why you were out here?"

Toni leaned over, her fists balled on her thighs. She was out of breath.

"It's complicated."

"Oh." Jess wasn't sure she had time for Toni's problems right now. "Everything OK?"

"Not really. But look, tell me what's happening. What can I do?"

"Toni!"

Jess looked past her friend to see a pale figure running towards them: Roisin Murray.

Toni turned. "Get back inside, Roi. Your mum can't see you out here."

"I don't care about my mum." She fell on Toni and threw her arms around her. Her face was pale, reminding Jess of how she'd looked when they got her out of her cell at the kidnappers' farm.

"Roisin," she said. "You OK?"

"I will be." Roisin stared into Toni's eyes. "I'm sorry. I shouldn't have let my mum tell me what to do. She's got no idea."

Toni kissed her briefly, then looked towards where Roisin had come from. "Does she know where you are?"

"I snuck out."

Toni nodded. Her lips were tight and her face pale. Jess heard sounds behind them.

"Damn," she said. "Come on, you two, we need to sound the alarm."

"Why?" asked Roisin. She shrank into Toni's side.

"There are people here from Filey. We need to tell everyone to get indoors."

"No you bloody don't," said Toni. "We fight back."

"Toni, please…"

Toni pushed Jess out of the way and watched the advancing crowd. She turned to Jess. "It's not like you to run and hide."

"I'm not saying run and hide," said Jess. "But they've got weapons. We aren't prepared. We should sit it out, wait until morning."

"What, and let them loot and vandalise our village, and God knows what else? Sod that."

Toni grabbed Roisin's shoulder. "Can you bring me the garden spade, from the cupboard in my kitchen?"

"What are you keeping garden tools in your flat for?" asked Jess.

Toni eyed her. "I never knew when I might need to defend myself."

"This isn't about you. It's about those idiots, and about the council trying to take our houses."

"All the more reason to stand up to them. Jess, don't worry. I'll knock some people up. It'll be fine."

Roisin was sobbing. "I'm not fetching it."

Toni wrapped an arm around her. "Please, Roi. I'm no use without a weapon."

"No, Toni. Please. Just come home with me. Keep out of it."

"Home? Back to your mum's, where I'm despised?"

Roisin blushed. "No. Back to your flat."

"And what will your mum do if she finds you there?"

"We can deal with that when we come to it."

Toni kissed her again. "That's very brave of you. But I have to help. We've got to fight. They aren't doing this to us again." She pushed Roisin gently away. Roisin wiped her cheek and started running towards the flats.

Around them, doors were opening and people were emerging.

"Get inside!" Jess shouted. "It isn't safe."

No one paid any attention. A group of people was forming behind her. The invaders had stopped moving now. They were in a line, between Jess and her house.

She ran to the growing crowd of villagers. Colin was at the head of them.

"This isn't like you," she said.

"They can't do this."

"We can't fight them," she said. "I've seen them up close. They've got better weapons than us."

"We're more determined than they are."

She shook her head. This was going nowhere. "Keep

this lot under control, will you?" she told him. "I'm going to find Zack."

13

MARTIN STOOD at the door to the street, his chest rising and falling. Sarah had her hand on his arm, trying to hold him back.

"I don't understand," she said. "You didn't want to go to the village meeting. Why do you want to get involved in this?"

He turned to her. "Don't you get it?"

"No. I really don't." She scratched her arm; it itched, the skin raw.

"This is my fault. I have to help make it right."

"How is it your fault?"

"This all started with me and Robert."

"Robert would have found someone else to do his dirty work."

He shrugged her hand off. His skin was hot, like it might set her on fire. "But he didn't," he said. "He found me."

"If he hadn't persuaded you to do it, you wouldn't be here now. With me. Have you ever thought about that?"

His face softened. "You really see it like that?"

She wasn't sure how she saw it. Mostly, she tried not to think about the circumstances of their meeting.

"You think it was a good thing I did, because it brought us together?" he asked.

"No. Of course not. But it's complicated, isn't it?"

He leaned in and kissed her forehead. "I love the way you see the good in everything."

"Thank you. So don't go."

"I have to. I'm one of you now, I have to prove myself. I can't let outsiders take over this village."

"You're talking like you've lived here all your life."

"Not yet, I haven't." His eyes were boring into hers; she felt a shiver run down her spine. She didn't dare think about the future. When her father had been at home, the future represented getting through the next day without him losing his temper. Now the future was a looming wall she couldn't imagine scaling.

"You'll get hurt."

"People are already getting hurt. You think I should be any different?"

He was right, of course. If there ever was logic in getting involved in a fight, his was solid enough. She grabbed his arm. "Keep safe," she muttered.

He pulled her to him and kissed her, long and hard. She savoured the taste of his lips, the smell of his skin, hoping she would feel this again.

She pushed him away. "Go. Before I change my mind."

Behind him, two people ran past. She couldn't tell if they were from the village, or not. She felt her eyes widen, her skin bristle. "Be safe," she whispered.

"I will. You go inside. You're too fragile to be out there."

She put her hands on her hips. "I'm a lot less fragile

than I look. And my mum's out there. I'm going to check on her."

"Please, Sarah..."

"You can't tell me what to do. That's what my father did."

"It's not safe."

"She needs me. I'll come with you as far as the corner, then I'll head to her house. I can take side routes, I know my way."

"I don't want you risking it."

She took a breath. "You don't tell me what to do."

"You know that's not what I..."

"Whatever. I'm going."

He opened his mouth as if to speak, then shifted his weight, seeming to deflate. She knew what he was thinking.

"Be careful," he muttered.

"I will."

She gripped his hand and together they ran the first hundred yards. Up ahead, torches blazed and men were shouting. They passed a woman who was slumped over another figure, talking incoherently. Sarah shuddered.

"Wait." She approached the woman. "What's happened?"

The woman looked up at her. She was bent over a man, unconscious on the ground.

Sarah fell to the ground and put a hand on the man's back. There was no sign of injury. The man shifted under her hand, his back moving against her skin.

"What happened to him?" She looked up at the woman; she didn't know her. She was as sure as she could be that this woman didn't have children. Had they ever met?

The woman said nothing.

"I'll get help," Sarah said. "I'll find Ruth."

The woman stared at her, her eyes wide. Then she looked at Martin. Her eyes narrowed and she shook her head. "He'll be fine."

"He's not moving."

"I can deal with it."

"But that's ridiculous."

The woman moved forward to grab Sarah's arm. Sarah winced; her grip was tight. "Why won't you let me help?"

The woman stiffened. "Go, please. We'll be fine."

Martin whispered in Sarah's ear. "Your mum, remember."

She waved a hand in his direction. "This is more important."

"Please," said the woman. "We'll be fine. You have somewhere you need to be, clearly."

She felt her chest tighten. She looked into the woman's face, trying to determine the reason for her mistrust.

She remembered her mother. Alone, in that cottage. "I'll see if I can find someone from the council. My mum." She shook off the woman's hand.

The woman nodded.

"Please, Sarah," said Martin. "I don't want you out here any longer than you need to be. I'll find Ruth."

The woman was watching them, not bothering to disguise her interest. The man on the ground was sitting up now, rubbing his forehead. Sarah moved away from them, towards Martin. She lowered her voice.

"I said, don't tell me what to do."

"Sorry. But please. I love you, Sarah. I don't want you to get hurt."

"I know." She forced out a smile then stood on her tiptoes and kissed him lightly on the cheek. She took one last glance at the woman, who was leaning into the man, pulling him up to standing. Her mother would know what to do.

She pulled away and ran towards a narrow gap in the houses opposite. These cut-throughs were dark, and difficult to spot. She'd be safe.

14

THE ROUTE to Jess's house was blocked. Someone had pushed a tangle of fallen branches into the alleyway that led to it, and in the darkness she'd failed to find a way through. She headed back for the village square, where villagers and attackers were tearing into each other.

The attackers had weapons: basic ones, but enough. Baseball bats, a scythe, the glint of knives. They wore heavy, dark clothing, bundled up against the night.

Some of the villagers were in pyjamas; a few had thrown coats over their day clothes. One or two had knives but others fought with kitchen utensils. This was not going to end well.

Jess ran towards two young men who she'd taught as teenagers. They were fighting three strangers, men with hoods hiding their faces. She pushed through the crowd, feeling the air stir as a garden fork passed her head.

"Stop it! Stop it, now! We all have to get inside!"

The young men ignored her. Instead, they pushed the hooded men aside and waded into the crowd of torch-bearing attackers, wielding garden tools. Once of them held a hoe and the other a garden fork and a hammer. Jess felt herself blanch at the thought of what damage it might inflict.

There was a crash off to one side and a flash of light from behind the houses close to her. She ran towards it, not stopping to consider if this was wise.

She stopped in the village square, panting. Ahead of her, the JP was ablaze, flames pouring out of an open window. She stared at it, unable to move.

Clyde ran out of the main door, shouting.

"Help! We need water!"

Jess ran towards him. "There are buckets in the boathouse."

"That's too far away. We don't have time."

She tried to focus. The flames were thickening; she could feel heat in the air. People ran towards them: Colin and three other men.

She span round. "Where can we find buckets?"

"The village hall, the storage room," replied Colin.

She frowned. "That's just school supplies."

"The space in the roof. It's full of junk. There are a dozen buckets up there, God knows why."

People were already running towards the village hall, Colin behind them. She sped after them; she had a key. Colin caught up.

"It's alright," he called to her. "I can get in. You go and check on people. Make sure no one's hurt."

She stared then nodded at him, her mind awash. Running away, she could hear sounds behind her; shouting and the crackle of the flames. Someone was chanting something. Were more people coming out to confront them, or would they stay safe in their houses? Could she blame them, if they did fight back?

But if they won this, if they repelled the invaders, the

authorities would never forgive them. Anita Chopra would be back here in the morning, court order miraculously expedited, ready to evict them all. And worse.

She turned to see the torchlight brighten behind her as it rounded the houses. It was accompanied by a crowd, at least twenty-strong. She felt her stomach sink.

Ahead of her, a group of villagers approached, young men employed at the earthworks. Was Zack with them?

She ran towards them. "Stop!" she screamed. "Can't you see this is what they want?"

But no one could hear her. Between the roaring of the torches, the shouts that came from all round, and the whistling of the wind, her voice was drowned out. She clenched her fists at her sides, summoning more energy.

"Stop!" she yelled.

One of them turned towards her: Sam, Zack's brother. "Sorry, Jess!" he called. "We can't let this happen."

She stared at him, her arms wide. He shrugged and carried on his way, picking up pace. She had to get back to the JP, to check it was empty. But the way was blocked by fighting.

The two crowds met in the middle of the village square. They blurred into one. The noise rose: shouts, and screams. The clash of weapons.

She collapsed to the ground. "Stop, everyone. Just *stop*."

This wasn't going to work. Diplomacy, politics, negotiation. None of it was worth anything in the face of the blind rage she could see on the faces around her.

A man came out of the throng and shoved past her, blood oozing from a cut on his forehead. She knew him; Mark Palfrey, Sally's fiancé.

Where was Ruth? People would need her help.

She took a side route to Ben and Ruth's house. Ben

would be out here somewhere; not like him to avoid a fight. But Ruth; Ruth would be in there, with her boys.

She hammered on the door. No answer. She knocked again. She put her face to the door and tried to shout. But there was smoke in her lungs, and her voice was hoarse.

This was useless. If Ruth was in there, alone with her boys, there was no way she would come to the door.

Jess ran back to the village square. Their attackers consisted of twenty, maybe thirty men. They carried torches and weapons. Axes, hammers, knives. Her own side, the villagers, numbered more, but had been caught unawares. Some of them were attempting to fight with their bare hands. Others wielded kitchen knives, or garden tools.

This had to stop.

She waded into the crowd, throwing her arms out in an attempt to protect herself.

"Jess! What the fuck are you doing?" Ben grabbed her by the arm and dragged her out. She stumbled then managed to catch herself, thumping his arm to push him off. He held a length of metal piping; his eyes were bright.

"You have to stop!" she yelled at him. "Can't you see, we'll lose this either way!"

"We have to defend ourselves."

He disappeared back into the crowd. She stared after him, her heart pounding in her ears. She could feel something warm on her cheek. She raised a hand and looked at it to see blood on her fingers. Hers, or someone else's? She licked it, feeling as if she was watching herself licking her own bloody fingers; not inside her body but somewhere outside, observing from a safe distance.

"Arrrrgh!" A man threw himself at her. She pushed him off, her fingers clumsy against the rough wool of his coat. He was one of hers.

"It's me, Jess! Don't hurt me!"

It was dark, and windy. The sea pounded in her ears. She was confused, and scared. She didn't have a weapon. These men were just as terrified: who knew who they might stab?

"Stop," she croaked. "You all have to stop."

They could retreat to their homes. The houses were solidly built, with brick or stone walls and heavy windows to keep out the cold. They would be safe. Anita Chopra would have nothing to accuse them of.

But no one was listening.

She half-ran, half-fell from the crowd and towards the tree that sat on the corner of the road leading out of the village. This tree had been here before them, before this holiday village was built. It would outlast them.

She staggered into it and let herself slide to the ground, glad of the solidity of the wood at her back.

She watched the crowd morph and sway in front of her. Blood dripped into her eyes; where had that come from? She blinked, and felt the world blur in front of her.

She wasn't up to this. The steward's role was to protect the village, and she'd failed. She didn't have the mental strength to convince them what was right, or the physical strength to stop this.

She'd failed.

A man ran past. Large, nimble, but heavily built.

"Zack, stop!"

He hadn't heard her.

She pushed herself upright. "Zack, wait!"

If she could persuade him to stop, then the others might follow his lead. Zack loved her. He was going to marry her. They'd talked about the future, about children. The thought terrified her, but she was prepared to give in to it. It was the only way to build a future, after all.

She thought of Ruth and Ben's wedding, out there on the beach five years earlier. It had rained. Everyone had been relieved when the ceremony was over, and they could retreat to the JP.

Would the whole village turn out for her wedding, as they had for Ben and Ruth? Or would they shun her? The woman who couldn't protect them. The steward who couldn't keep their children safe.

She had to find him. She stumbled forward. "Zack!"

A shape at the edge of the crowd stopped moving. Was it him? She couldn't tell. Her eyes hurt.

She picked up her pace. She had to get to him before he melted into the crowd.

She was almost on him when she heard the shout; loud, high-pitched, terrible.

"Zack!"

She started running. The shape she'd been aiming for swayed and fell to the ground. Maybe it wasn't him. Surely it wasn't him. It couldn't be him. They were getting married.

She almost tripped over a mound on the tarmac.

"Zack?"

She fell to the ground. It was him, covered in blood.

She raised her face to the night and screamed.

15

Dawn's house was dark, the curtains closed against the night. Sarah shifted from foot to foot as she waited for her mother to come to the door, glancing around to check no one had followed her.

She leaned against the door and knocked again, listening for movement inside. There was nothing.

She looked up and down Dawn's road. Dawn lived in a semi-detached cottage, near the Meadows. Not far from Martin's flat. The front window was dark, no sign of movement or flicker of a candle.

Was her mother out there, in all this? She felt her chest sink.

She looked back towards the centre of the village.

"Mum!" she called. "Mum! Dawn!"

She caught movement from the corner of her eye. In a window opposite, a curtain moved.

Don't shout, she told herself. *Don't panic.*

She ran back towards the centre of the village. She could see people moving in the village square. Torches lit the shapes that advanced on each other then receded. Someone screamed and she threw a hand to her chest.

Mum, where are you?

She started towards the fighting, wondering if this was the right thing to do. Her mother wouldn't be here, surely? Dawn knew when to stay away from trouble.

Martin's flat was off to the left, closer now than her mother's house. Maybe Dawn had been in there all along, but just too afraid to open the door. That would be the sensible thing to do.

Someone was running towards her. She ducked into the shadows, unsure if it was a villager or a stranger. She watched as a woman sped past.

"Mum?"

The woman stopped. "Sarah?"

She stepped out of the shadows. "What are you doing?"

"Helping."

"Helping with what? Not fighting, surely?"

"Don't be silly. I was with Pam. We were making sure the village hall was properly locked, and checking the shop."

Sarah didn't know her mother was friendly with Pam. "Why?"

"Because we don't want anyone getting in there, of course. There's a lot of food in that shop. If it was stolen, we'd go hungry."

"Can't someone else do that?"

"No, love. Someone had to step up."

"I think you should be at home."

Dawn raised an eyebrow. "Do you now? Is that why you're out here, running around like a banshee?"

"I was looking for you. I wanted to check you were safe."

"I'm fine, sweetheart. You need to be indoors."

"It's alright. I'm going back to Martin."

Dawn clutched her upper arm. "He's not in the middle of all that then?"

"I talked him out of it."

"He'd be vulnerable."

Sarah swallowed the lump in her throat. Behind her, she heard a shout, and glass shattering.

"Go home to your young man," Dawn said. "He needs you, tonight."

"I'll walk you home first."

"I'm perfectly fine."

"I don't mind."

The eyebrow went up again. "Sarah. I can look after myself."

"I'm worried about you."

Dawn flapped a hand. "I've faced worse than this."

Sarah tightened her jaw. She didn't know if Dawn was referring to the years of domestic abuse, or to what had happened on the walk here from Somerset, after the floods. To the young men Ted had killed, defending his daughter's honour.

"Go," Dawn said. "Make sure he's alright. I wouldn't want him getting hurt, for your sake."

Sarah kissed her mother gently on the forehead – something she'd never done when their father was around – and started running.

16

JESS SLUMPED TO THE GROUND, her skin tight. Zack had fallen against her, his weight heavy on her legs.

She looked up.

"Help! Someone, help!"

But there was too much noise. A man ran past, maybe fifty yards away, oblivious to her.

She leaned over Zack, pawing at him, pulling him.

"Zack. Zack, talk to me! Are you hurt?"

Zack said nothing. His eyes were blinking, his face pale, only his left side visible.

She tugged at him, wondering why he'd never felt this heavy before. His jacket felt matted with something.

As she turned him on his back, his weight caught and he continued moving until his face was against her stomach.

There was a knife in his neck. Its dark handle stared at her.

She grabbed it and tugged. A spurt of blood covered her fingers, making her yelp. She threw the knife to one side and clutched his neck, pushing against a wound she couldn't see for the blood.

She looked up again. People were running towards her. She had no idea if they were villagers, or attackers. No torches.

"Help! Help me!"

Someone broke free of the group and sprinted towards her. He almost crashed into her, then clattered to the ground, kneeling in front of her.

"Ben! Where's Ruth?"

Ben looked from her face down at Zack. "What happened?"

"He's been stabbed."

"What?"

"Where's Ruth?"

"She's at home."

"Jess. Zack. Oh my— Zack!" Sam was behind Ben, blocking out what little light there was. He pushed Ben to one side. "Get Ruth. I'm taking him to the pharmacy."

Sam crouched and placed his arms beneath his brother. Zack grunted. That was good, wasn't it? At least he was making sounds.

Sam groaned and lifted Zack, hauling him onto his shoulder. Jess stared at him, her chest full of ice.

Sam started towards the pharmacy. His steps were slow and uneven, but he wasn't going to drop Zack. Jess could be sure of that. She threw her hands over her face and peered through her fingers, as if blocking it out could make it go away, could undo the last few minutes.

"You OK, sis?"

She nodded. "Ruth," she croaked.

"Yes. Course." Ben put a hand on her shoulder. "Are you OK to get to the pharmacy? I'm sure I can find someone…"

She shook her head, her eyes closed. "No. Yes. Just go. Get Ruth. I'll make it."

Ben squeezed her shoulder and ran off towards his house.

Jess pulled herself up to standing, ignoring the damp that soaked her jeans. Around her people were running, shouting. She heard a cry. A woman.

Stop it, she whispered. *Just stop it, all of you.*

Sam was out of sight now, the darkness having swallowed him up. She forced herself to take a step forward, swaying as she moved. She had to go with him. Zack needed her.

She clenched her fists, then pummelled them against her thighs. Her mind felt full and empty at the same time, like she was floating in a vacuum.

She wiped her eyes and looked at her fist. It was smeared with blood. She put her fingers on her eyebrow; it stung.

It was nothing.

"Zack! Sam!" she cried, and made for the pharmacy.

17

RUTH RAN TO THE PHARMACY, Ben calling after her.

"Stay!" She looked round to see him standing in the doorway. "Stay with the boys!"

When Ben had got home, she'd been lying on the floor between the boys' bed, staring at the blank ceiling. She had no idea how long she'd been there. Ben had almost battered the door down to rouse her.

Ahead of her, shapes loomed out of the darkness. Someone shouted, off to her right. She felt her hair shift as someone ran past. She turned; who was it? But they were gone.

She set her jaw and focused on the route to the pharmacy. It may be dark, with the streets blocked, but she walked this route every day of her life.

The village shop had its door hanging open. She grabbed the doorframe and pulled herself in, preparing herself for what she might find inside.

Jess was in the doorway to the pharmacy, slumped against the doorframe, her body shaking. Ruth put a hand on her arm.

"You OK?"

"I'm fine." Jess wiped her face. Her hand came away bloody.

Jess nodded towards the table in the centre of the space. Sam stood over it, his back to Ruth, blocking her view. Behind him she could see the lower half of Zack's body. The lights were on; someone had made an exception to the energy conservation rules.

"Sam. Excuse me, please."

He turned to face her. His face was pale, streaked with dirt, eyes dark and wide, his pupils huge. He seemed to have lost the six years she'd seen him grow since arriving at the village, despite his bulk.

"Help him," he said.

She nodded. "I will."

Sam shifted to one side, his eyes on Zack. She slid past him to inspect her patient.

Zack's feet hung over the edge of the table closest to her. He wore jeans and a heavy coat. Blood soaked the collar of the coat and covered his neck. His face was pale and his eyes closed.

She felt her stomach shift. She stared at his neck, at the blood. It pulsed once, twice. She held her breath.

"We need to get his coat off." Her voice was thin and low.

Sam stepped forward and started taking his brother's coat off. He was gentler than she expected.

"Here." She took the sleeve, the one on the opposite side to the wound. Jess stepped in. Together, they managed to remove the coat without disturbing Zack too much. Ruth stared at the fabric of the coat the whole time, not wanting to look at the wound.

The coat dropped to the floor. Jess grabbed Zack's hand.

"I'm here, sweetie. Ruth's here. She's going to make you better."

Ruth turned to stare at Jess, horrified. This was impossible. She turned back to Zack, her stomach full of lead.

Zack's eyelids flickered, just slightly. Ruth gulped down the bile that was rising in her throat.

She blinked. Sweat was pouring into her eyes. She wiped the back of her sleeve across her face and muttered to herself: *you can do this.*

She opened her eyes to see Robert Cope in front of her, lying on the table. A knife in his throat, blood pouring onto the kitchen floor.

She heard a whimper.

"Ruth?" Jess's voice was sharp in her ear. She shook her head to banish it.

She squeezed her eyes shut then looked back down, prising her eyes open as slowly as she could. Zack's inert figure lay on the table, his eyes closed. She felt a long breath flicker through her lips.

"Oh thank God," she breathed.

"What? Is he going to be alright?" Jess sounded shocked.

"I don't know."

Ruth kept her eyes on Zack's face. If she looked away again, he would be replaced by Robert. She heard Robert's voice in her ear: *Mrs Dyer*. She shuddered.

"Leave me alone!"

"Ruth? What's going on?"

Jess pulled at Ruth's shoulder. Ruth stared at Zack: she mustn't take her eyes off him. She had to treat him, try and save him. She could do this.

"Sam, grab me a pair of gloves. Up there, on the shelf."

Sam fumbled with the gloves and got them into Ruth's

hand. She put them on blindly, staring at Zack's closed eyes. His face was turning from white to grey.

"Stethoscope," she said, pointing towards the hook where she knew it was hanging. The cold metal landed in her palm and she placed the earpieces in her ears without once taking her eyes off Zack.

She placed the chest piece on his chest, fumbling with the buttons on his shirt. Nothing. She moved the instrument around, desperately searching for a pulse. She raised a hand for silence. Sam was breathing heavily beside her, Jess's breath fluttering in her ear.

The door clattered open. "Zack! Oh my Zachary!"

A tall woman with her hair in a severe ponytail pushed past them and threw herself over Zack. Sam grabbed her.

"Mum, leave him. Let Ruth do her job."

Ruth shook her head. Zack's face looked cold and small now, nothing like the strong face that had so charmed Jess. She heard Jess sob behind her.

"I'm so sorry. There was nothing I could do."

She allowed her eyes to travel away from Zack's face and down to his wound. She should clean it. She couldn't face that.

"I'm sorry," she repeated. She'd taken her eyes off his face; she couldn't risk looking again.

She turned to Jess. Their eyes tore into each other, Jess's searching her own for some sign that she was wrong, that she might change her mind.

She shook her head, the motion frantic, and pushed past Jess and into the shop. She stumbled towards the counter, almost hitting her head on it as she fell to the ground. She leaned around it and vomited onto Pam's clean floor.

18

FOG HAD DESCENDED over the village, casting the houses in a dense shroud. Sarah picked her way along the streets, aware of how easy it would be to miss a turning.

The streets were quiet now, with the shouting and yells from earlier having died down. Up ahead, something was on fire. She approached it hesitantly, afraid there might still be people there. It was surrounded by villagers, a chain of people with buckets attempting to put out a fire in the composting bin at the back of the JP. The stench of burning plastic rent the air, making her gag.

She scanned the chain of people for Martin. No sign of him. Was he somewhere else, still fighting? Or had he gone home? Maybe he'd been told to go home, that this wasn't his fight.

She swallowed. When would Martin be made to feel he belonged here?

She'd wanted to stay with Dawn, to spend the night in the cottage she'd made so homely. In the house Dawn had shared with Ted, there had been no adornments. Only the crucifix on the wall. But here, the house was full of objects.

Drawings Sarah had done as a young girl, smoothed-over pebbles from the beach, anything that was shiny or beautiful. She no longer had to worry about any of these objects being used as a weapon against her.

Go home to your young man, Dawn had said. *I wouldn't want him getting hurt, for your sake.*

But what about Dawn's sake?

Now she was on the main road which ran from the village centre to the outside world, parallel to the coast. Buildings loomed at her through the mist. She counted them. Theirs – Martin's – was the third block of flats on the left.

The building was still and dark, no candles flickering at the windows. She turned her key in the lock and slipped inside, her ears alert for the sounds of people. Martin might be back, or their neighbours might be inside, listening. She knew the two of them were an object of interest.

She hurried up the stairs, her breath short, and knocked gently on Martin's door. She rubbed her hands together against the cold and waited.

No answer. She tried again, but he wasn't in.

She pulled her keys back out of her pocket. She didn't like letting herself in when he was out, it still felt like trespassing. But she wasn't about to go back to Dawn, not in this fog.

She pushed the door open and blinked as a draught hit her face. They would never have left the window open, not on a night like this.

She crossed to it, muttering in irritation. She reached out for the handle but it was already closed. She froze.

In the middle pane was a jagged hole, not much bigger than her fist. She stared at it, then walked to it and touched the rough edge with her fingertip.

Had the attackers come this far? Had they been

breaking windows? She hadn't seen any evidence of that on any of the other buildings she'd passed.

She stepped back, aware of the crunch of glass beneath her feet. Hundreds of tiny shards littered the carpet. It would be impossible cleaning them all up in the dark. She'd have to warn Martin, tell him to steer clear until the morning.

She stepped backwards, shaking out her feet. Then she saw it. At the edge of the glass, a dark object. She frowned and bent to it. Anxious not to lose her balance.

She picked it up between her fingertips, carefully in case it had glass stuck to it. It was hard, and rectangular. A brick.

A length of string was wrapped around it, and beneath that was a piece of paper.

She almost dropped it in her shock. She glanced up at the door, wondering again where Martin was, if he was safe.

She retreated to the kitchen and placed the brick on the worktop. She reached into a drawer and lit a candle, taking care not to bring it too close to the paper. She unravelled the string and placed it carefully in her pocket. Good string couldn't go to waste.

She glanced at the door again, wishing she'd closed it. Then she unfolded the paper. She squinted to make out the rough black letters in the candlelight.

Leave, or die.

19

Ruth headed for the clifftop, half running, half staggering. Her chest felt like it might burst and her stomach ached. Her legs were weak and a sharp pain worked its way up her body, threatening to overwhelm her.

Her mind was thick with images. Robert's face, in that cell. The bedroom he'd locked her in. The vase of flowers she'd knocked to the floor. Then there were the smells. His aftershave, heavy and cloying. The institutional smell of the cooking in that kitchen. The toilet she'd been forced to sling her waste into, from a bucket.

She collapsed to the ground, not sure if she'd reached the beach or was still on the clifftop. Or was she on the path between the two? She couldn't tell. It wasn't dark now; her eyes were filled with light, green-filtered light like she'd had in that cell with its algae-encrusted window.

She let herself slide to the ground, resting her head on the grass next to the path. The sky was dull, only a few stars visible now. She thought she could remember being brought down here, brought to the boat. Thrown over someone's shoulder: was it Robert's, or Martin's?

That was wrong. She'd woken up in her cell, not knowing where she was or who had taken her. She had remembered nothing from falling asleep in her son's room, to waking up almost two days later.

But she could see it. She could feel the rhythm of his steps beneath her, the sensation of her body shifting as he carried her. She could hear the voices, men muttering to each other. Congratulating each other.

She let out a cry. Was she remembering, or imagining? Was her mind filling in the blanks? And why couldn't she get it to stop?

Above, a star flickered. There was a break in the fog cover, just big enough to make out a patch of clear sky. It felt bright, brighter than the sun, so bright she might burn.

She rolled over, whimpering into the grass. She shoved her hands into the ground and pushed herself up. She didn't know why, but she had to get to the sea.

She stumbled down the path, not stopping until her feet were slowed by sand. The fog had thickened now and she could see nothing; not the stars, not the sea, not the waves that lashed at her feet.

They'd brought her here, and taken her. She'd come back this way, after – after…

Robert's head flashed in front of her eyes. The way his eyes had bulged as she'd twisted her foot into his chest. The way the blood had eased its way out of his lips, like he was regurgitating red wine. She felt her insides hollow out. She needed the toilet.

She reached behind a rock and yanked her trousers down. She relieved herself noisily, tears rolling down her face. She pulled her trousers back up, not caring about cleaning herself.

She crawled towards the water. If she went out there, if she went back to the farm, would it all go away? Maybe if she saw for herself, could know for sure that he was dead. She hadn't even seen them bury him.

"Ruth!"

She wailed. Sheila Barker was behind her, her hands looping under Ruth's armpits. She dragged her up, repeating her name.

"Ruth, are you alright?" her neighbour's voice was loud.

"Leave me. Please."

"Ruth, please. I'll take you home. It's not safe out here."

She felt a shudder rip through her body. She turned to see Sheila looking down at her, eyes full of worry.

What was she doing? Why was she here, kneeling at the edge of the sea? She couldn't even swim.

"Oh my God."

"It's alright, Ruth sweetheart. They've gone now."

"What? Who?"

"The attackers. They've gone."

Ruth frowned. "Don't tell Ben."

"Don't tell Ben what, sweetheart?"

"This. About this."

Sheila smiled. "Don't worry. I'll take you home."

20

"What's that?"

Sarah crumpled the note between her fingers and pushed it into the palm of her hand. Her skin was damp and hot. She could feel heat rising to her cheeks.

"A brick. It got thrown through the window. Did you see anything?"

Martin closed the door to the flat behind him. His face was smudged with dirt, his hair dishevelled.

"Are you alright?" Sarah asked. "You didn't get hurt?"

"I'm fine. No one could see who I was in the dark so they let me help."

"How is it?"

"Bad."

"Tell me."

He took a step towards her. "We managed to fight them off, but not without..." his shoulders rose and fell. "Not without casualties."

Sarah thought of her mother, of her pale face as Sarah had run off into the night. She shouldn't have left her.

"Who?"

He put a hand on her shoulder. "Who?" she repeated.

"Zack. He was one of the people who came to the—"

"I know who he is," she snapped. "Sorry."

"Sam's brother," Martin whispered.

Sarah nodded. "How is he?"

"Not good. They took him to the pharmacy."

She nodded again.

"You were reading something," he said.

"What?" She crunched the note even tighter in her hand.

"You had something in your hand. Was it from your mum?"

"No. Why would it be—?"

He put his hand out. "You're pale."

"I'm fine."

"If you don't want to show me, that's fine. I respect your—"

"It's not that."

"What, then?" He pulled a hand through his hair, catching on the knots. He wiped his cheek; a clear patch appeared.

"You need a shower," she said.

"There's not enough water till the morning. I'll have a wash. I tried checking on your mum's cottage, Sarah. But there was no answer."

"I told her not to answer the door."

"When?"

"After you left. I went to see her, remember?"

His brow furrowed. "Yes. Sorry."

She sat down on the sofa, pulling him down next to her. "Was it awful?"

He nodded.

"Are they really gone?"

"For now."

She swallowed the lump that had risen in her throat. As she lifted her hand to wipe her eyes, the note fell onto her lap.

She looked down at it. The word *die* was clearly visible.

"What's that?" Martin asked. "Something about dying." He looked from the note to her face. She closed her eyes. She didn't want him to know about it; but did she have the right to hide it from him? To lie to him?

There'd been enough lies.

She picked it up and unfolded it. She held it out, her fingers loose.

He took it, his eyes not leaving hers. He brushed her face with his fingertips then lowered his gaze to the note.

She watched his eyes travel quickly across the page, then the look of shock register on his face.

"I'm sorry," she said.

"Don't be." He stared at the note. "When?"

"It was here when I got back from Mum's."

He looked behind him at the window. He shrugged off her hand and walked to it. He fingered the edges of the hole as if that would tell him something.

"Wait here." He hurried out of the door and clattered down the stairs. She followed him, lingering in the doorway. Whoever it was might still be out there.

She stumbled down the stairs. "Martin! Come back! What if they're there?"

Outside, he was standing on the grass at the front of the building, staring up at the window. He bent to examine the ground, peering around him. He ruffled through the neatly tended shrubs that separated the grassed area from the road.

"It might not be for you," she said.

"*Leave, or die*. It's not for you."

"You don't know."

"Sarah, you've lived here since the beginning. I've been here six months. It's for me."

"But who?"

He pursed his lips. "I don't know. It could be anyone. It could be one of the women we took."

"Ruth?"

"No. Not Ruth. And she was in the pharmacy, anyway."

"Ben?"

He said nothing.

"You think it was Ben?" she asked. Again he said nothing.

After a few more moments staring up at the window, he shivered. "It can't have been. He's Jess's brother. If he felt that strongly, I wouldn't be here."

"I don't think he and Jess get on."

He shook his head, then looked up and down the road as if expecting someone to appear out of the shadows. "Let's get inside."

She held out her hand and he took it. Together they closed the door and retreated upstairs. They passed the other three flats in the block and she stared at the doors, suspicious. Who was it who hated them this much that they were prepared to send death threats?

In the flat, Martin went to the sink and gulped down a glass of water. He refilled it and emptied it again.

"You're not safe here," he said.

"I'm fine."

He turned and slumped against the sink. "I'm putting you at risk."

"You don't know it was for you."

"Sarah," he cried. "You're being naive. Of course it was for me!"

"Maybe."

"Someone wants me dead. I don't want you getting caught up in it."

"I'm not leaving you alone."

She leaned against him. He gripped the sink behind him, not touching her. His breathing was ragged. "You have to, Sarah."

"I'm not leaving you."

He pushed her off and held her at arm's length. "You know I love you. You know I don't want to push you away. But you can't be hurt because of me, not again. Please, I'll take you to your mum's. You should sleep there."

"You can come too."

"I'm not putting your mum in danger."

Sarah couldn't argue with that. "Let me stay here."

"Please, Sarah. I'll lock the door. I'll barricade the window. But I don't want you here."

"No."

"What if they take it out on your mum? Who's to say she hasn't had a brick through *her* window?"

"I'd know."

"How?"

She felt her stomach tighten. "I would."

"Not in the middle of everything that's been going on tonight. No one would know. Go and check on her, please. Stay with her. We can work out what we're going to do in the morning."

"I don't see what we can do."

He slid down the front of the kitchen units coming to a stop with his legs splayed on the floor in front of him. "I'll think of something. D'you want me to walk with you?"

She thought of her mother, alone in that cottage. Peering through the curtains, watching the fighting outside. Would it remind her of Ted? Would she stand too close to the window?

"I'll go on my own," she said. "I'm not scared of this village."

He stood up. "Please, let me."

"No." She walked to the door, then remembered herself and returned to him. She kissed him on the chin. He bowed his head to return the kiss and they stood there, locked together, for a few seconds. Tears ran down her face.

"I'll see you tomorrow," she said. "Stay safe."

As she left the flat, she didn't turn back. Looking at him again would mean she'd never manage to get out.

21

Jess stared at the deep trench that had been dug before the sun even came up. It ran alongside the one they had dug for Sonia five years ago.

Sonia's grave was covered in lush grass now, wildflowers in a chipped vase at its head. The vase was new; the last one had been lost to the weather. Jess had been meaning to create something more permanent, to plant flowers, but the position on the exposed land to the north of the village meant she'd never been able to persuade anything to grow.

The wind whipped at them today. It tugged at Jess's hair, threatening to pull her off the hillside and over the cliffs. Ahead of her stood Zack's family, opposite the trench. Leah was leaning on her husband Tim, her eyes dry but her body overcome with shaking. Next to her Zack's three younger sisters. Behind them Sam, his face blotched with red and white, a cut which snaked down his right cheek. His clothes were dirty and dishevelled and he looked like he hadn't been home since the previous night.

Jess didn't know where Sam had been when Zack was

stabbed. She didn't know if he'd been close, or if he was off somewhere else fighting another group. She wished they'd been together. They would have protected each other.

She felt something push up from the pit of her stomach into her throat, like a gust of wind or an almighty heave of breath. She turned to the side, expecting to be sick, but nothing came.

Zack's body lay beside the trench, not more than a metre from Jess's feet. He was wrapped in a greying sheet. He looked smaller than he had in life.

She'd stayed with him for as long as she could, hunched over the table in Ruth's pharmacy. No one had seen Ruth since. She'd caught a glimpse of Ben on her way here this morning but he'd been hurrying somewhere and had ignored her when she called out to him.

Clyde stood next to her. He'd helped Tim and Sam carry Zack here, Colin trailing behind. Colin stood a few metres away, a respectful distance. Jess wished he would get closer. They all needed to gather together now, to support each other. Everyone here had suffered loss in one form or another.

She felt her legs weaken, and clenched her feet inside her boots in an effort to keep her balance. She was the steward; she had to maintain control. And the family opposite her had more right to be grieving than she.

They all stared at Zack's body in silence. She wondered what, or who, they were waiting for. Ben? Ruth? The rest of the council?

No. At Sonia's funeral, there'd only been five people. Herself, Ruth and Ben, Clyde who'd dug the grave, and Murray, the steward at the time. He'd conducted the ceremony.

Her head snapped up. They were waiting for her.

She opened her mouth to speak, then felt a wave of nausea. This was impossible. Just twelve hours ago, she'd been laughing with Zack. He'd told her a funny story about something Sam had said to him about weasels. She felt her eyes well up.

She looked at Colin. He was staring down at his feet, his face pale.

"Colin?"

As the council secretary, he was the next most senior person here.

Colin looked up. "You need me to do the honours?"

She nodded, grateful that he understood.

Colin stepped forward. Leah looked up at him, then at Jess. A dark shadow crossed over her face, then dropped as Sam clutched her elbow.

Colin took a breath. "We are here this morning to bury our son, brother and friend, Zack."

And fiancé, Jess thought. But Colin didn't even know they were engaged. Only his family did. And hers.

Where were her family? Had Ben been hurt, and she didn't even know it?

No. Ben had gone home to watch the boys, when Ruth had been summoned out for Zack.

She turned and scanned the route back to the village, wondering who knew they were here. Should they have told more people, given them the opportunity to come?

But this was how they did things here. There were no undertakers, no coffins, no embalming. Just the elements to take the dead back to the earth. Burials were done quickly so that the body could be safely in the ground and not presenting an infection risk.

Colin was looking at her. Not just Colin: Leah and Tim too. Sam gave her a sad smile over his mother's head.

"Sorry," she said.

Leah shook her head. Sam nodded at Jess. "Are you alright?" he asked.

She felt uncomfortable. She looked at Leah and Tim, their swollen eyes. How would it feel, to lose a child?

"Yes," she muttered. "Sorry. Carry on."

Colin licked his lips. "Zack was a valuable and productive member of the village. He earned funds for his neighbours by going out to work outside our community, and he was always a willing volunteer when we needed someone to undertake a dangerous or difficult task."

Leah sobbed. Jess felt tears wash her cheeks. She listened to Colin's low voice, pierced by Leah's sobs, as he remembered what the man she loved had meant to the rest of them.

22

Jess's feet felt numb as she tried to pick her way across the rutted field. Behind her, a mound of earth marked the spot where the man she had been supposed to marry now lay. Ahead of her, his parents walked, huddled together.

Above them, seagulls wheeled, shrieking at them from a pale blue sky. From time to time their cries would be accompanied by a moan from Leah.

The custom after any ceremony at the village was to go to the JP. Weddings, namings, funerals: all of them were marked by the community coming together afterwards. She hoped more people would be there, that she wouldn't have to face the onslaught of Leah's raw grief alone.

They reached the path back to the village and her feet began to feel secure. Colin caught up with her and fell into step.

"Thanks," she said.

"Of course," he replied.

Colin would be thinking of his brother, who died on the journey here. He, like so many others, hadn't been afforded the dignity of a funeral. At least Sonia had managed to make it here, to spend her last months in peace and warmth. At least she hadn't died violently at the hands of other refugees, or of people reluctant to welcome refugees to their homes.

"Do you think they'll come back?" she said.

"Hmm?"

"The people from last night. They'll be back, I'm sure of it."

Colin sighed. Could she take it, if they were besieged again? Could she stand to sit inside and watch the violence, or – even worse – to go out there and play her part, without Zack next to her?

"I don't know," Colin said.

But he did. They all did. No one was safe anymore.

They rounded the first pair of houses, identical beach houses that had once been designed to provide a touch of New England glamour for holiday makers, but whose wooden frames were now disintegrating. They had never tried to house anyone in these structures; the storms six years ago had battered them, exposing the short-sightedness of the construction company in not making everything out of brick or stone. Suddenly the story of the three little pigs came to mind and she stifled a laugh.

"You alright?" Colin asked.

"Yes. Sorry."

"I know how it can hit you. Sometimes it can feel like your reactions are... inappropriate."

"But I can't be inappropriate in front of his parents."

Colin looked ahead to where the Golder family had disappeared around a corner. "It can't be easy."

"We were getting married."

Colin stopped walking. "That was quick."

"Six months. Not that quick."

"No. I'm so sorry, Jess."

She sniffed, wishing she'd thought to bring a handkerchief. But she'd spent only a few hours at home since the fighting had stopped.

"Thank you."

They rounded the corner where the Golders had disappeared, and almost crashed into them.

Jess grunted. Why hadn't they carried on walking? Why weren't they in the JP?

Then she saw. Up ahead of them, outside the back of the JP. Two police cars, and two others. One long and dark, the other small and blue.

She pushed through the grieving family, ignoring Leah's trembling hand on her shoulder, and called out.

"Not today! You can't be here today!"

She ran towards the cars. She peered inside the first one; it was empty. She ran between them all, pulling on the handles, shouting through the windows. But no one was inside.

She heard a bang behind her. She turned to see the door to Ben and Ruth's house closing. Ben was outside, marching towards her.

"Where were you?" she breathed. "We just buried Zack."

"What are they doing here?" he hissed. "We said no!"

"Have they spoken to you?"

He marched past her and slapped the side of one of the police cars. He turned and looked round the square. Behind her, Colin, Clyde and the Golder family stared. She could sense their breath being drawn in.

"Ben, please. We're in the middle of a funeral."

"Where are you? Why have you come here again?" Ben was ignoring her, pushing her out of the way in his haste to reach the centre of the village. He stared wildly around him as he moved, as if expecting someone to appear in the shadow of the houses.

She followed him. "Did you see them arrive?"

He didn't reply, but carried on walking. She kept pace, her legs heavy.

They passed the JP. Either side of them, people emerged from their front doors, asking questions. Ahead was Sarah Evans. She'd been walking away from them, towards Martin's flat. She looked scared.

"Jess? What's happened?"

"Have you seen any police come past?"

Sarah blanched. "No."

"We're here."

A woman emerged from a side road.

"You! What are you doing here?" Ben ran at her.

"Ben, stop!" Jess called. They couldn't afford to antagonise her.

"Miss Chopra," she said. "What are you doing here? Why so many cars?"

"We've brought you some new residents."

"But we didn't agree."

"We don't need you to agree. Read this."

Anita pulled a grey folder out of her battered brown leather bag. She opened it and pulled a sheet of paper out, handing it to Jess.

Jess scanned it, the words dancing in front of her. *Courts - order - mandatory*.

"I don't get it."

Ben snatched the paper off her and read it. He threw it to the floor.

"It says the council can move residents in and out of the village at any time, if the housing need is sufficiently acute."

"Do you know what they did to us?" Ben yelled. "Have you heard what happened last night?"

"You gave as good as you got, I hear." Anita Chopra had her arms folded across her chest. Jess realised this woman had never been on her side. She'd never intended to negotiate.

"But we've been here six years!" Jess cried. "Surely that counts for something."

"Sorry."

Jess felt her chest hollow out. She took a step towards Anita. The other woman stared back at her, her face calm.

How dare you, Jess thought. Heat rose from her stomach into her chest. She felt sick.

She felt her arm come up, not aware of having decided to lift it. Her fist clenched. She gritted her teeth and thumped Anita, catching her with a glancing blow to the jaw.

Anita stumbled back, her eyes wide. She glared at Jess and clutched her face. Jess stared back at her, her heartbeat deafening in her ears.

"I just buried my fiancé," Jess said. "Surely you could have waited."

23

Sarah stared at Jess, her mouth falling open.

Had she really seen the steward punch someone?

The woman was sitting on the grass. A man stood behind her, writing in a notebook. DS Bryce, from Filey police station. She eyed him, hoping he wouldn't start asking questions about Martin.

Jess stood over the woman. Her body was tense, quivering almost. Colin Barker put a hand on her arm.

"Jess?" Sarah whispered. "Are you alright?" Stupid question.

Jess turned to her. Her face was pale and her features hard. "Zack's dead."

"I know." The Golders were standing behind Jess. Sam looked at Sarah, his face soft. His sister was sobbing into her mother's shirt.

Sarah smiled at Leah Golder, who gave her a tight nod.

"I'm so sorry," Sarah said. Holding her voice steady was a struggle. Other people were going through worse than she was.

"What about me?" The woman on the ground was

dabbing her chin with a grubby handkerchief. There was blood, but not a lot of it.

The woman stood up. She raised a finger, was about to jab it into Jess's chest. She stopped herself.

"You can't do this. I have every right."

"Don't you know what happened here last night?" Jess replied.

"I imagine you're about to enlighten me."

Ben stepped forward. "People from Filey. They attacked us. My sister's fiancé is dead."

The woman frowned. "What people from Filey?"

DS Bryce cleared his throat. "Are you saying he was murdered?"

"Yes." Leah Golder. Her voice was thin. She put an arm around Jess, who sank into her. "My son. He was going to marry Jess. You robbed us of his future."

"I didn't do anything," the woman said. Sarah knew she was from outside, that she was an authority figure. Jess had talked about her at the village meeting.

"We'll need to speak to witnesses," said DS Bryce. Leah glared at him.

"Bit late for that."

"I was there," Jess said. The woman was fingering her nose. It didn't look broken. Sarah wondered if she should offer to find Ruth.

"What did you see?" asked the detective.

"Nothing," Jess sighed. "It was dark. Chaotic."

"Maybe someone else will have..."

"No one will talk to you," Ben interjected. Jess frowned at him. But he was right. The police; the council; the attackers. They were all the same, as far as the villagers were concerned.

"Do you want to report a crime?" DS Bryce said.

"Yes," said Jess.

"No," said Leah. She looked at Tim, her husband. He nodded agreement.

DS Bryce shook his head. He plunged his notebook back into the inside pocket of his jacket.

"This doesn't change anything," the woman said to Jess. "We're still appropriating those houses. I could report a crime myself. Assault."

Jess stared at her. Her chest was rising and falling; would she do it again?

"But I won't," the woman continued. "Just leave me alone to do my job. Or you'll find yourself under arrest."

24

Ruth stared at the sea, letting its ebb and flow fill her mind. She focused on the waves, approaching and receding; the hiss and crash of the water as it hit the rocks to her left. The sea washed over them, splashing her every few seconds. She was wet, and cold.

She didn't care.

She focused on the horizon, trying to drag the memories back, to remember what it had been like to be taken out there. Somehow, if she could remember being taken, if she could recall the trip to the farm on the tiny village boat, she would feel better. She had to know that there was nothing she could have done, that she didn't bring it all on herself.

But her mind was blank. From the moment she'd lain down on the boys' floor, to walking up in that cell, there was nothing.

Every time she closed her eyes, she saw his face. Robert Cope, smiling at her. Threatening her. Touching her. *Mrs Dyer*, he'd called her, as if her marriage was more important than she was. To him, it turned out, it was. If it hadn't been for Ben, for that stupid thing he'd done as a teenager, none of this would have happened. He'd told Ruth nothing about it, not until the confrontation in the farmhouse kitchen, when Robert had accused him.

Ruth had to believe Ben had told the truth. She had to believe he'd never meant to kill that man, that it had been an accident, and really Robert's fault. Of course it was Robert's fault; he was the type. But it was Ben who had struck the fatal blow. Ben who had followed Robert blindly like he had no mind of his own. Ben who'd lied to the police.

She stared at the grey water, willing the images out of her mind. She'd woken this morning to find a spray of wild flowers on the kitchen table. Had she put them there, or had Ben? An image of the cornflowers Robert had picked for her sprang to mind, falling to the floor as she'd thrown them off the mantelpiece in his bedroom.

Then there was the feel of fabric against her skin. Everything she wore was the shirt he'd made her wear. The stiff white cotton, neatly pressed and tighter than a straightjacket around her.

She drew her lower lip into her mouth and bit down, hard. The pain helped to bring her back to the here and now. Maybe she should find something sharp, something she could hurt herself with, to remember where she was. That would be something Robert hadn't done to her, a new sensation.

Still biting down, she pushed herself up to standing. Her boys had been asleep when she'd left the house. Ben was out somewhere; he'd rushed out when she was sitting at the kitchen table trying to keep her breakfast down. She'd sat in the silence of the house for one hour, maybe two, not moving, hardly daring to breathe. The day was still young; she'd woken just as dawn was sending its tendrils around the bedroom curtains. Unable to lie next to Ben and listen to him breathing, she'd crept downstairs, and felt herself slump when he'd come down right after her.

Eventually, she couldn't take it anymore. She'd stood up and walked out of the house without even taking her coat off the peg. It had started to rain: a hard, cold rain that seemed to bite her through the skin. She didn't care. The cold was good, it reminded her she was alive.

But her boys had been alone for over an hour now. She had to go back. She pulled in a breath. The thought of going through the motions, of making their breakfast, sending them off to school, of going to the pharmacy, filled her with lead.

The pharmacy. She couldn't go back there. She couldn't even remember if she'd returned to clean up after Zack's death. She'd rushed out, unable to contain herself, desperate not to intrude on the Golders' grief. If she hadn't, would someone else have cleaned up? Jess, maybe?

People would understand if she didn't go in to work today. But then, they'd need her. Zack wouldn't be the only one injured. They might be waiting for her already, lining up in the shop under Pam's hard stare.

She had to do it. Duty could pull her through the day. And her love for her children.

She dragged her feet though the sand – was she still wearing slippers? – and headed for the village. She could do this.

At the top of the cliff path, she slowed. She couldn't bear for anyone to see her like this.

The roads were quiet, the rain sending everyone indoors. The houses were dark, people inside not looking through their windows she hoped. Nevertheless, she kept to the shadows.

"Ruth?"

She looked up to see a woman approaching her. She took a step back, horrified.

"Ruth are you alright? It's just me, Sheila. I'm with Benji, he needed the toilet."

Ruth heard the snuffling of Sheila's dog in the undergrowth. She withdrew from it, scared of the dog suddenly coming at her.

She said nothing.

"I'm on my way home, if you'd like to walk with me," Sheila said. She was wearing a long raincoat and a hat. She didn't comment on Ruth's attire. This wasn't the first time they'd met out here; was Sheila watching her?

Ruth couldn't refuse. Sheila lived right next door, with her husband Colin. If she told the other woman she wasn't heading home, then Sheila would want to know where she was going. And why she was soaked through.

"Thanks," she muttered. She fell into step with Sheila, who talked to her dog as they walked and didn't ask questions. Ruth shivered as she trudged towards the houses, every muscle in her body on edge.

They came to her house first and Sheila stopped, grabbing the dog by the scruff of the neck. He was a small dog, a mongrel she'd found lost on the cliffs just over a year ago. He was tubby and spoiled, the darling of the children Sheila taught.

"Thanks," whispered Ruth. She scuttled towards her front door, not looking back.

She clattered through the door and shook herself off. *Please, let the boys be safe upstairs*, she thought.

She froze. The boys were sitting at the kitchen table, Ben between them. Ollie sniffed out a sob when he saw her and Sean pushed his chair back and started to approach her.

"No," said Ben. He put a hand on his son's arm. Sean's face crumpled and he sat down in his chair.

"Finish your breakfast for Daddy."

Sean nodded and picked up his fork. Eggs and bread. Was that what she would have given them, if she'd been here?

Ben sidled towards her. "What were you thinking?" he hissed.

"I wasn't long. I needed some air."

"I got back here and found them alone. Anything could have happened to them."

"You weren't here either."

"I left them with you. Not alone."

She ran her tongue across her lips; they were calloused. "You ran out without saying anything. I was looking for you."

"I was with Jess. Zack's funeral."

She felt herself crumple. "I'm sorry."

She pushed down the wave of emotion rising from her stomach. She wanted to lash out at Ben, to pummel his chest until he bled. She wanted to shout at her boys, to demand why they hadn't covered for her.

How could she think about her family like this? Her family, who loved her?

She met Ben's gaze, unable to think of anything to say. After a few long moments, she felt a whimper escape her lips.

She couldn't let them see how she was feeling. She couldn't let it go.

She ran up the stairs and slammed the bedroom door behind her, throwing herself onto the bed.

25

"We can't take them."

"We don't have much choice, Ben."

Jess pushed her fingertips into her forehead, fighting off a headache. She'd convened a meeting of the village council, to discuss how they were going to respond to the actions of Anita Chopra. Not to mention the attack last night. She wondered how many of them had been talking about her punching Anita. There had been more than one muttered *well done* or *you told her* when they arrived.

"Come on," said Toni. "Let's not argue among ourselves. We have to work out a plan. Where are the people who've been evicted going to live?"

Ben turned to Jess, facing away from Toni. "They go back to their homes."

"And then what happens to the families they've put in those houses?" Jess asked.

"Not our problem."

Jess sat back in her chair. She didn't need this. She hadn't even had a chance to speak to Leah Golder.

"Look," she sighed. "I'm in that big house that used to be Dawn's." She ignored Dawn's blushes. "It's just me, with three bedrooms. I'll move out, and they can have my house."

"But where will you live?" asked Dawn. She looked agitated, as if expecting to be sent back to the house where she'd suffered Ted's cruelty.

Jess turned to her brother. "Ben?"

He frowned and scrunched up his mouth. "If you leave your house, you're just saying they've won."

"I know, Ben. And I don't want them to kick us out of our homes any more than you do. But there are nine people currently sitting in the JP surrounded by all their belongings. They need a place to go."

"You can come in with me," said Toni.

"Really?"

"Yes. I've got one of the flats. It's only one bedroom, but I'll take the sofa."

"No, I will."

"Whatever."

Jess considered. Ben and Ruth had a spare room. But they also had a house with two young boys in, that felt a lot fuller than it really was. Toni was out most of the time, and wouldn't expect much of her.

"OK. Thanks, Toni. I'll go and tell them."

"Wait," said Colin.

Jess wanted to sleep. She wanted to lie down on her bed – or Toni's sofa, it seemed – and remember Zack. She wanted to process her guilt about not conceding to Anita Chopra's demands. If she had, the attack would never have happened.

"What?" she replied. Colin glared at her. "Sorry, Colin. Is there something else?"

"The attack. Who's to say it won't happen again

tonight? We need to agree how we're going to defend ourselves."

"I think we'll be safe," said Dawn.

"What makes you think that?" snapped Ben. Jess shot him a look.

Dawn sniffed. "They've won, haven't they? When someone thinks they've won, there's not much point in attacking again."

The room fell silent, no doubt each person processing their own understanding of Dawn's experience with aggressive behaviour.

"We can't assume that, though," Colin said. "We need to be ready."

"Too right," muttered Ben.

Jess looked past him and out of the window. It was growing dark, the April evening descending. They shouldn't stay here any longer than they needed to.

"Any suggestions?" she asked.

"Quite a few people have weapons," suggested Clyde.

Jess shook her head. "That's not the answer. We need to keep our people safe, not let things escalate. I saw what happened last night. I saw how useful our so-called weapons were."

Clyde dipped his head. "Sorry, I didn't mean to…"

"It's alright."

"We have to defend ourselves," said Ben.

"No, Ben," said Jess. "We have to keep our people safe."

He glared at her. "How about we use those newcomers as human shields?"

"Ben! You don't mean that."

"If they think they might be attacking their own, maybe they'll be put off."

"That's barbaric. I'll pretend you didn't say it."

She looked round the others. Dawn was staring at Ben, her cheeks pale. Dawn's only prior contact with Ben was when he had fallen apart after Ruth's disappearance. She was used to men like Ted. Bitter, aggressive men. Maybe that was what Ben was turning into.

Jess shook herself out. Ben was no Ted. She saw him with his sons every day. He loved them more than she'd ever seen a man love anyone.

"I suggest we batten down the hatches," she said.

"I agree," said Toni.

"You think that'll work?" asked Colin.

"I don't know what will and won't work," Jess replied. "But these houses are solidly built. And they have had a small victory. I don't think they'll be as desperate this time. We should tell everyone to stay safe, to stay indoors."

"We need to post guards, at least," said Sanjeev. Ben threw him a nod. Jess looked between Sanjeev and Toni. Each of she and Ben had their own ally on this council; why did it all have to be so confrontational?

"OK. Any volunteers?" If she was giving up her house, she wasn't about to give up her night as well.

"I'll go," said Toni. "Give you some space."

"Thanks. Anyone else?"

"Me," said Sanjeev.

"Don't you need to stay with Navi?"

"My brother is eighteen now. He can take care of himself."

Jess thought of Sanjeev and his brother, who also lived in one of the large houses on the clifftop. Maybe they should volunteer to share, as well.

She wasn't about to start forcing people out of their homes. Sanjeev's parents and sister had died in a fire after the floods; they'd suffered enough.

"OK," she sighed. "But two people isn't enough."

"I'm pretty sure Martin will do it," suggested Dawn.

"Martin?"

"He wants to prove himself."

"Makes sense. Anyone else?" She looked at Ben.

"I'll go," said Clyde. He threw Jess a smile that reminded her of the way he used to flirt with her, before she'd grown close to Zack.

"Thanks." She stood up. "Right, it'll be getting dark soon, and we've got things to do. You all know the cascade system. I'll cover my share. Just make sure everyone knows to stay indoors, and bolt their doors. OK?"

26

THE SCHOOL HAD BEEN quiet today. All but five of the children had been kept home by their parents, scared after the attacks of the previous night.

But Sheila had insisted that they owed it to those children who had turned up to provide them with a semblance of normality. She'd tasked Sarah with running a history lesson. Sarah had forced herself through it, impatient for the day to end so she could talk to Martin.

Coming out of the village hall after tidying up, she saw Dawn going past with a group of council members. Toni, Sanjeev and Clyde, people Dawn had never spoken to until six months ago. Now they were her friends. Sarah felt a rush of pride.

"Mum!"

Dawn paused in her conversation. She put a hand on Toni's arm and said something to her, then hurried across the damp grass to her daughter.

"Hello, love. Everything alright?"

Sarah hadn't told Dawn about the note. She'd explained her agitation by saying that she and Martin had had an argument. *Don't worry, love*, Dawn had said. *It's perfectly normal.* After an awkward pause, Dawn had asked if Martin had hurt her. Sarah had bit back the urge to snap that not all men were like that, but managed to keep it inside.

She'd spent a restless night staring at the grey ceiling of her bedroom, listening to her mother's snoring through the wall. She'd never noticed it before. After finally falling asleep at three am, she'd slept late and had to rush straight to the school when she woke. So she hadn't seen Martin since he'd told her to leave.

"I'm fine," she said. "Have you been at a council meeting?"

"We needed to agree what to do about the new people."

"What new people?"

"You haven't heard? They've forced us to take in some people from Filey."

"They're here already?"

"They got two of the cottages next to the Meadows. The Argyrises and Snellings were forced out, told they weren't using their homes to full capacity."

Sarah frowned. "Where will they go?"

"Jess is going to move out."

Sarah felt a shiver run down her back. "Where will…?"

"Toni's."

"Right." Sarah barely knew Toni. She'd gone south with Jess to rescue her and the others, but her focus had been on Roisin Murray, not Sarah. "I need to speak to Martin," she said. "Will you be alright without me tonight?"

"They're imposing a curfew."

"A what?"

"We're all to stay indoors once it gets dark. Apart from

the people who'll be keeping guard. I thought Martin might volunteer."

"You did?"

"He's always saying how he wants to play his part. Helping to protect us is a good way to do that, don't you think?"

Sarah thought of the note. *Leave, or die.* Could helping to guard the village convince them that Martin belonged here now?

"I'll tell him."

"Thanks. He's to assemble at the JP, as soon as it's dark. Which will be soon. You come straight home after you've seen him, yes?"

Sarah looked into her mother's eyes. They were pale, and hooded. Was she scared?

"Of course."

The two of them hurried towards Dawn's house. The sky was showing the first signs of dusk. Sarah insisted on walking Dawn home and when she'd heard her mother bolt the door behind her, she went to Martin's flat.

The outside door was locked. That was unusual, but then these were unusual times. She supposed one of the neighbours had learned about the curfew. She fished in the pocket of her skirt for her key and slid the door open as quietly as she could, making sure to lock it behind her.

At the top of the stairs, Martin's door was open. He was hunched over, scrubbing the floor just inside the doorway.

"What are you doing?" she asked.

He looked up. His hair was greasy and he looked like he'd barely slept. He wore a pair of pink rubber gloves; she had no idea where they'd come from.

"What does it look like I'm doing?"

"OK." Why was he being so tetchy? "Are you scrubbing the floor?"

He blinked at her then returned to his task, his hair bobbing as he moved.

"You're scrubbing the floor, at nine o'clock at night."

He didn't look up. "Yes."

She felt cold. She looked at the carpet beneath him. It was shaded by his body. "Why?"

He stopped scrubbing. After a moment's hesitation, he looked up at her. His face was pale.

"Because someone thought it would be fun to post a pile of shit through my letterbox."

27

Jess stared at the row of cottages. Four of them huddled into each other, their backs to the Meadows, the field to the south of the village which would have been for walking dogs back when this was a holiday village. Now, it was a wasteland. To one edge sat the rubbish dump for the village, and at the other was the pit into which they dug any rubbish that couldn't be reused or recycled. They'd learned to be creative with their waste: having no access to the disposable conveniences of the world before the floods had concentrated minds, and now the amount of waste that went into that pit each month was less than she had thrown out in just one bin bag every week.

The two houses closest had been appropriated for the newcomers. She had no idea why Anita Chopra had chosen these: perhaps there was some logic behind it, perhaps it was random. These certainly weren't the only houses the authorities would deem under-occupied. There was her own for starters, now the home of the two couples who'd been turfed out of these two houses, and plenty of other two-bedroom cottages with only two adults living in them. The flats had one bedroom but there were only sixteen of those in total. Holidaymakers liked large houses, with light and space and sea views.

If Anita Chopra was planning on continuing with her quest to rehome all the people affected by that explosion, they'd all be doubling up. It would be like the journey here all over again. Like the spell they had spent in Leeds at her Aunt Liz's house. Six adults in one two-bedroom flat. No wonder Liz and her wife Val hadn't complained too much when they said they were heading north.

Standing here would do no one any good. They were probably watching her from behind the mesh curtains that adorned the windows of the two cottages. Or if they weren't, their neighbours beyond would be. Watching her, discussing the way she'd betrayed them all.

She walked to the first door and knocked on it, trying to project confidence. She heard movement behind the door.

"Who is it?"

"My name's Jess. I'm the village steward. I've come to welcome you, help you settle in."

There was muttering behind the wood. She tried to remember which of the two families had been given this house: was it the Asian family with their three kids, or the white couple and their hulking teenage sons? She wondered if those boys would be prepared to go to the earthworks with the other lads and earn money for their new community.

The thought of it made her think of Zack. She fought down a wave of emotion and focused on the door knocker. It was brass, and shaped like a lion. Dark patches of dirt surrounded it.

"Hello?" she called, standing closer to the door. She didn't want the neighbours listening in.

"We don't need settling in, thanks."

"I just want to help you feel more at home in your new house."

"We won't be staying for long. You can leave us now."

"I don't want to intrude."

"You already are."

More muttering.

"One question."

"Go on," she said.

"The electricity isn't working."

She frowned. From what she'd been told, electricity was rationed in Filey as much as it was here. Maybe that was a lie, designed to keep the villagers from wanting to move there.

"It only works for an hour a day. The first hour after sunset."

"How am I supposed to know when that is?"

She looked at the sky. It was darkening, the pale white clouds turning grey at the edges.

"It should be in the next fifteen minutes or so."

"What time?"

"I don't have a watch. Sorry."

"How do you know?"

"I just do. You learn. Can you let me in?"

"No. Leave us alone."

She stared at the door for a moment longer, aware that the person – woman? – she'd been speaking to would be watching her through the peephole. Then she nodded and gave a small, awkward wave.

"OK. You can find me in the first block of flats on the Parade. Or I'm normally out and about around the village. Ask anyone for Jess, they can usually find me."

"That won't be necessary."

She resisted the urge to pull a face at the door. "OK.

Well, I hope you have a pleasant evening."

No response. She backed up the path and went to the next door.

This time, it was opened immediately. A plump woman in a headscarf stood in the doorway, one child in her arms and another leaning against her leg. She gave Jess a nervous smile.

"You're the steward."

"Yes. How did you know?"

The woman looked towards the house Jess had been outside before. "They don't like us."

"Sorry?"

"It's just us, and them, and all you lot. But they hate us just as much as you do."

"We don't hate you. Whatever made you—"

The woman shook her head. "Who lived here before?"

"Mary and Hugo Snelling. They're an elderly couple."

"And now they're homeless, because of us. Of course you hate us."

"They aren't homeless. They're living in my house now."

"So where are you?"

"With a friend."

Why was she telling the woman all this?

"Look," she said. "This village took us all in after the floods. That's what happens here. You'll be made to feel welcome, once things have died down."

A man appeared behind the woman.

"Aabida. I told you not to answer the door."

The woman turned to her husband. "She's friendly enough."

"How do you know she's genuine?"

Jess extended a hand. "I'm sorry. They didn't tell me your names. I'll Jess Dyer. You are…?"

"Mohammed Bagri," replied the man. "This is my wife, Aabida."

Jess gave him a smile. He didn't drop his frown. "Pleased to meet you. I'm the village steward. If you have any questions or concerns, please come and find me. Or if anyone behaves inappropriately towards you."

"You're trying to tell me you'd take our side, over that of your own people?" the man said.

"It's not like that."

He grunted. He tapped his wife on the shoulder and ushered her back into the house.

"We'll be keeping our doors locked tonight," he said. "We didn't ask to be sent here."

Jess nodded. Keeping their doors locked was a good idea; the rest of the village would be doing the same thing. "You don't know if anyone from Filey is planning to come here tonight, do you?"

"We had nothing to do with that."

"I'm not saying you did. But if you heard anything, you'd tell me, I hope. Your family would be just as much at risk as anyone else."

Another grunt. "Unless we daub our house in blood. Like the Jews at Passover."

Jess had no idea what he was talking about. "Please don't daub your house in anything. Just stay inside. You'll have electricity for an hour, starting very soon. Have you been given your food rations?"

"If you call this stuff food."

"It's the best we can do. You'll get used to it."

He sighed. "I guess we'll have to."

She nodded again. "Anyway. I need to get home, before it's dark. Don't forget, if you need anything, ask for Jess.

He closed the door without saying anything. Biting down her anger, she turned and made for Toni's flat.

28

"You have to report it," said Sarah.

"There's no point."

She stared at the dark patch on the doormat where he had cleaned it up. The smell was in her nostrils, sharp and hard. It felt as if it would never leave the flat.

"Who do you think it was?" she asked.

"Same person who threw that brick through the window."

She looked towards the window. He'd patched it up with a piece of rough wood while she'd been at the school. It didn't look as if it would withstand much.

"I hope so."

"Why?"

"Because if it isn't, that means there's two people victimising you."

"It could be anyone," he said. "They all want rid of me."

"No they don't. Stop talking like that."

He stared at her. "I'm putting you at risk."

When he'd first said this, it had been endearing. It had

shown how much he cared about her. But now, she was beginning to tire of his paranoia. If her mother had integrated herself into the community after all they'd been through, why couldn't Martin?

"Stop saying that. We just need to go to the council, that's all. Find out who it was, and make sure they're punished."

"You think they will be? How can you be sure it isn't someone off the council?" He twisted his lips into a sneer. "Ben, for one, hates me."

She said nothing; Ben would take longer than most to accept Martin's presence here. After all, Martin had accepted the man's hospitality and then stolen his wife away in the middle of the night.

"Look," she said. "Let's lock the door tonight. Cover up the letterbox. We can deal with this in the morning."

"I can't stand to keep the windows closed. It stinks."

"It stinks of vinegar more than anything."

"I can still smell it."

"I know. I can too." She looked at the dark patch again, and shuddered. Who would do such a thing? And was it human excrement, or from an animal? The village kept pigs, but it hadn't smelt like it came from them.

"We haven't got much choice," she said, trying to breathe through her mouth. "I'll go and see Mum, see if she can help."

"I don't want her to know."

"She's part of the council."

"I know." His brow was creased. "But it's so humiliating."

She wrapped her arms around him. "I know. I'm sorry."

They stood like that for a while. She tried to ignore the smell that engulfed the flat, but it was impossible. She had to tell someone. He couldn't let them treat him like this.

"Look," she said. "I told Mum I'd spend some time with her tonight. I won't be long."

"OK." He glanced towards the door. "Don't tell her about this. Not yet."

"She might be able to help."

"I just want to keep it private."

"Martin—"

"Please?"

She sighed. "Alright. Not tonight. But you can't keep this secret forever. What if they hurt you?"

He said nothing.

"You need to accept help." She put a hand on his arm. It was stiff.

"I'll lock the door. You go to your mum."

"I'll be back soon. Before lights-out."

"Be careful. Please."

"Sure." She kissed him on the cheek, then let him pull her to him and fold her in a hug. His body was tense, and he smelled of vinegar.

She was careful to avoid the damp patch as she left, and crept down the stairs. There was a chance that one of their neighbours had done this.

She hesitated at the bottom of the stairs. Maybe she could sit outside, watch them? If anything else arrived at their flat and no one entered or left, then she would know it was one of them. Or she might see them.

No. Nothing else would happen tonight. They'd made their point.

She needed help. She understood why Martin wanted to keep this quiet, why he didn't want to draw attention to himself, but she knew there was one person she could trust.

She ran to her old house, where Jess lived now. The

house looked as forbidding as it always had; Jess's presence hadn't affected the way it looked from the outside. She didn't know what she'd been expecting; Jess was too busy to change the house. But somehow she was disappointed that as she walked up the path, she felt like she was coming home instead of visiting the home of a friend, the familiar dread rising in her chest.

She knocked on the door and a young woman she didn't recognise opened it.

"Yes?" The woman looked tired. Her hands were damp and she wore a threadbare apron Sarah recognised as Dawn's.

"Oh. Is Jess here?"

"She moved out." The woman wiped her hands on the apron. "Just until we get our place back. You looking for her?"

"I don't understand." She didn't know why, but the idea of strangers moving into this house annoyed her somehow.

"They kicked us out of our cottage." *Of course*, thought Sarah. Dawn had told her; she hadn't expected it to be so quick. "Me and George," the woman continued, "and the Snellings too. Jess said this place was big enough for all of us. It is, too. Want to come in? I'll give you the tour."

Sarah felt her skin prickle. Did this woman not know that she had lived here until six months ago? Had she been that anonymous?

"I just need to find Jess."

"She said if anyone came looking for her, she'd be at Toni's."

All Sarah knew was that Toni lived in one of the flats. "Do you know where that is?"

"Hang on." The woman disappeared then returned with a slip of paper. She grinned at Sarah. "I know. Colin gave it to me. I haven't used paper in years. Anyway, it says thirty-six on here. So I guess that's the number of the flat. Does that help?"

Not much, thought Sarah. But she smiled and thanked the woman anyway.

Flat thirty-six was in the block next to Martin's. She approached it carefully, not wanting Martin to see her. It had turned dark now, and lights shone from the windows of the ground floor flat. She examined the numbers on the front of the building, and compared them to the numbers on Martin's in her head. By her reckoning, the flat with the lights on would be Toni's.

She pushed the outer door. It was locked. She felt herself deflate. Of course it was locked; they were all subject to curfew in about thirty minutes. She tried the doorbell, but knew it would long since have stopped working. Then she rapped on the wood. She waited. No one came.

She went to the edge of the illuminated window to her right. If there was someone inside, they might not react kindly to her face appearing at their window. But she had no choice. She peered round, hoping the growing darkness would give her some camouflage.

Toni was in the kitchenette beyond the living room, washing dishes. Jess was nowhere to be seen.

She pulled back, disappointed. Jess was probably out somewhere, doing steward work. Maybe if she went looking around the village…

That was pointless. This village was a warren of roads and alleyways. She'd never find her.

"Sarah?"

She opened her eyes to see Jess standing at the door next to her.

"Are you looking for me?"

She nodded, aware that she was blushing. "Sorry."

"Don't be." Jess glanced towards the lit window. "Walk with me."

They started towards the allotments. Jess walked fast and Sarah struggled to keep up.

"Everything OK?" asked Jess.

"Yes. I mean no."

Jess didn't break stride. "Go on."

"It's not easy to talk about."

"You can tell me anything, you know." Jess slowed momentarily, and looked at Sarah. Then she picked up the pace again. Her voice was low, as was Sarah's. Sarah was relieved that Jess no more wanted them to be overhead than she did.

"Is your mum OK?" Jess asked.

"It's not my mum."

They were at the allotment. Jess led her to a bench at its edge. Clay pots stood in a precarious pile at its feet. The bench had been patched up a few times over the years, and the wood of its back didn't match its base. Sarah lowered herself onto it. She scanned the allotment, hoping no one was around.

"They'll all have gone home," said Jess. "Don't worry."

"Right."

"So are you going to tell me what's going on? I assume this is related to Martin?"

Jess was smiling, her voice encouraging. Sarah stared back at her: what was the steward expecting her to tell her?

"You can't tell him I told you."

Jess frowned. The life fell from her face. "Why not?"

"Because I told him I wouldn't."

"Is that wise?"

Sarah said nothing. She stared at a bush ahead of them. It overflowed with small yellow flowers.

"Lying to him isn't a good idea. Not with your relationship having the history it does."

Again, Sarah said nothing.

"Sorry. I shouldn't have said that." Jess sighed. "Go on, tell me."

Sarah looked at her friend and mentor. Jess had puffy eyes, and her skin was pale. Her cheeks were clothed with red.

"I'm so sorry about Zack," she blurted.

Jess seemed to shrink, as if Sarah had hit her with a hammer. "Thank you," she muttered. She pulled her shoulders up. "Please, tell me what you need to. It's getting dark."

Sarah took a breath. "He's been getting death threats."

She felt the bench shift as Jess stiffened. "Death threats?"

"A note. And something... unpleasant... posted through the door."

"What kind of unpleasant?"

"Human excrement."

"Seriously?" Jess kneaded her fists into her thighs. "Sarah, I'm so sorry. Do you have any idea who it was?"

"No. It could be anyone."

"What makes you say that?"

"He's not exactly popular."

Jess didn't contradict her. Sarah appreciated that; she hated having to lie to Martin herself, to convince him that he was being accepted. Last night's attack had highlighted how isolated this community was, and she knew that plenty of people saw him as part of the problem.

"When did this happen?" asked Jess.

Sarah told her.

"You weren't at the flat?"

"No. The first time, I was with mum and the second, at

the school."

"Hmmm."

"Do you think he's making it up?" Sarah felt defensive.

"No." Jess shook her head. "No, that's not what I'm saying. But other people might."

Sarah felt cold. She needed to get back to Martin. God only knows what else might have happened while she was away.

"Can you help us?" she asked.

"Of course. That's my job." But Jess didn't sound convinced. "Look, you keep an eye out. See if you spot anyone watching your flat. If you get anything else, tell me."

"Will you ask around, see if you can find who did it?"

"I don't think that'll help."

"Why not?"

Jess raked her hands through her hair, picking at the tangles. "There are two hundred people in this village. Finding the person who did this won't be achieved by asking all of them, and hoping people tell the truth."

Sarah felt bile rise in her throat. "But what if they hurt him? Leave, or die, it said."

"People who do things like that rarely follow through."

"How do you know?"

Jess said nothing. "Come on. Let's get you home." She grabbed Sarah's hand and pulled her up from the bench. "I'll do everything I can. But I can't promise much. Meanwhile, stay vigilant."

"Of course."

"But I really don't think they'll carry through with it. Martin isn't at risk."

Sarah stared at the steward. Jess was always right. She always knew what to do, what was best for Sarah.

But this time, Sarah didn't believe her.

29

SARAH MADE her way back towards Martin's flat. The most direct route was across the Meadows and past the cottages at the southern edge of the village.

She hesitated at the edge of the allotment, peering into the darkness. *Stop it*, she told herself. She could skirt around the edge and be home in a few minutes. The dark didn't matter; she knew her way.

Nevertheless, she went faster than she normally would, hoping the extra speed wouldn't make her more likely to trip on the undergrowth or crash into a tree. Before long, she was at the gap in the hedge that led to the village centre.

She walked quickly but lightly along the middle of the road, not wanting to appear suspicious. If anyone was following her, they would have to show themselves if they wanted to approach.

There was someone up ahead; a man, overweight and balding. He leaned on the wall of a house. Was he smoking? No one smoked here.

She pulled back to watch him, ducking down behind a

bush. The door of the house opened. Candlelight spilled out. She could hear voices inside: loud conversation, and someone singing. She wondered which family lived there.

Two people approached the man. One of them brought something out of his pocket and then bent towards the first man's outstretched hand. A second cigarette sparked between them. She held her breath. They must be the newcomers, the people from Filey.

But how many of them were there? A family, Dawn had told her. With teenage sons.

She clenched her fists and forced herself to relax. The three men over there could very well be the father and his sons. She had nothing to worry about.

But she didn't move.

Their voices rose, loud in the still of the night. They were talking over each other, their words indecipherable. They had strong accents she struggled to follow.

The door to the house opened again. Inside were more people: at least four as far as she could tell. She frowned.

One of the people inside gestured towards the smoking men.

"Get back in here!" they hissed. "You don't want them seeing you."

Sarah felt cold sweat trickle down her back. She watched the men withdraw into the house. One of them leaned out of the door and peered up and down the road. Sarah crouched behind the hedge, as still as possible.

The man closed the door. She waited, wondering if they would be watching if she walked past their house. But she knew her way around this village. She ducked behind one of the houses opposite and sped around the back of it, running for Martin's flat.

30

"JUST WHEN I thought things might finally calm down around here," said Jess. She was sitting in a threadbare armchair in Toni's flat, sipping strong mint tea.

"I don't think this place ever gets easy," Toni replied. She was on the floor in front of the single log that sufficed for a fire, her arms wrapped around her knees.

"We had six years of peace, before I got lumbered with the steward job."

Toni turned to her. "Is that really how you'd describe it?"

Jess yawned. "OK, maybe not."

"There were attacks before, remember. And it was ages before the authorities left us alone."

"No one died then."

Toni stiffened. "I'm sorry, Jess. That was heartless of me."

Jess sighed. "No. I'm being self-obsessed. You were right."

Toni slid backwards along the pine boards to lean

against Jess's legs. The sensation was good; it made Jess feel grounded.

"I can't even mourn him properly."

Toni reached round to rub Jess's calf. "You can here."

"Not like his parents are doing. I missed the wake because of that bloody Anita Chopra. And they won't come near me. It's like I'm tainted."

"None of this is your fault."

There was a knock on the door. Toni flinched and drew away from Jess, inching back towards the fire.

"Come in," she said. Jess wondered who might be visiting this late.

The door inched open – Toni had insisted on not locking it, not until later – and a face appeared in the gap between the door and the frame.

"Tone?" came a whisper.

Toni stood up and threw Jess a nervous smile. "It's Roisin."

She opened the door and ushered the visitor in. Roisin started when she saw Jess. "Oh."

"It's OK, it's only Jess. She's cool." Toni had a hand on Roisin's cheek. She was staring into the other woman's face, her eyes dark. Jess remembered the way she had softened when they got Roisin out of her cell at the farm, traumatised and covered in blood.

Jess sat up straighter. "Hi, Roisin."

"Hi." Roisin looked from her to Toni, her expression puzzled.

"She's staying here for a bit," said Toni. "She gave her big house to the Snellings and the Argyrises."

"Oh. That was kind."

"I rattle around in that big old thing," said Jess. "I was happy to get out."

She thought of the last night she had spent there with

Zack. The night before the attack. He'd grilled some fish that he'd brought up from the smokehouse; extra rations, a perk of being one of the people who went out and earned money for the village. It had been burnt and bitter and she'd probably been unappreciative of his efforts. Now she remembered it as a delicacy.

"I'll go out for a bit," Jess said. "Leave you two in peace."

"It's OK, Jess," said Toni. She turned to Roisin. "How long have you got?"

"Mum thinks I've gone to drag Dad back from the JP. If he gets home before I do, I'll be in trouble."

Toni slumped. She'd told Jess about the problems she had gaining the acceptance of Roisin's family, when they were walking south to rescue her. The Murrays looked down on Toni, and kept pushing Roisin in the direction of Zack's cousin, John.

"It's fine," Jess said. "I need some fresh air anyway." And she needed to speak to Ben. After rescuing Ruth, she'd been under the impression she and her brother would be working as a team from now on; it seemed he had other ideas. She needed to bring him into line.

"You sure?" Toni wasn't looking at Jess, but at Roisin. Jess thought of Zack looking at her like that, and a pellet of something hard lodged in her stomach.

She nodded, unable to speak, and pushed past the two women.

Outside, the night was clear. Stars pulsed down from a dark sky, and she could hear the sea in the distance. Toni's flat was further away from the water than her own house, but not as far as the house she'd shared with Sonia, from which they'd very rarely heard the sea. But tonight she could not only hear the sea; she could smell it. Heavy and dank and salty, reminding her of the journey back here with Ruth. Of the rescue mission she'd made with Zack, when he was just another young man from the village.

She shouldn't be out here. The guards would be assembling soon and she had no way of knowing if they'd be enough. They could be invaded at any moment. She'd told the rest of the village to stay indoors, to lock themselves in, but here she was. Strolling down the Parade like it was a sunny afternoon.

Suddenly self-conscious, she slipped into the shadow of the houses to her left, and kept to the shadows as she made her way towards Ruth and Ben's. She hurried across the open space in front of the JP and peered inside the pub. It was empty, probably had been for some time. She could see Clyde inside, polishing the bar in the dark. She considered knocking on the window and giving him a wave but then thought better of it.

She crept past the pub windows and came into view of Ben's house. It stood in a row of almost identical houses, all of them built to the same design but with differences in the surface details. Ben and Ruth's, for example, was painted blue, while her own was clad in local stone. Sanjeev's house, between them, was surfaced in red brick. She wondered if it was built of brick, or if that was just cladding to hide featureless concrete.

As she prepared to step out of the shadow of the pub, she caught movement opposite. Ruth and Ben's door was opening. She pulled back and watched, wondering where Ben was going when there was a curfew on. She'd follow him; maybe she could talk to him more easily if Ruth wasn't in earshot.

But it wasn't her brother. Instead. Ruth slid backwards out of the door, peering into the house as if worried someone might see her. At last she drew away from the door and pulled it gently closed. She glanced up at the window above her and shrank into herself before walking away from Jess, towards the cliff path.

Jess stared at her, wondering if she should follow. Ruth had surprised her, creeping out of her own house like that. Sure, there was the curfew to consider, but it looked as if she was sneaking out behind Ben's back. Why?

Did she owe it to Ruth to go after her and help her, to warn her about the danger of attack? The first wave of invaders had come from the beach last night; who was to say they wouldn't come the same way again?

Or did she owe it to Ben to tell him what she'd seen? If she let Ben know that she'd seen Ruth leaving the house, then they could sort this out between themselves. But she'd already involved herself in their relationship more than she felt comfortable with.

Ruth was out of sight now. She was a grown woman; Jess had no place telling her what to do. She'd be back soon, surely.

She hurried across the road, peering in the direction Ruth had gone, and knocked quietly on the door. She turned the doorknob and pushed. It wasn't locked. Was no one observing this curfew?

She stepped into the dark house and closed the door behind her.

"Ben?"

A muffled sound came from upstairs. She stood at the bottom of the stairs and whispered her brother's name as loudly as she dare.

Ben appeared at the top of the stairs.

"Jess? What the fuck? You gave me a scare."

"Sorry. Your door was open."

"It was what?" He shuffled down the stairs and grabbed a key from its hook.

"Ruth's out there. She might not have her key."

"What?" Ben put his fingertips to his cheeks and pulled the skin down, dragging his eyes into long ovals. "What are you talking about? Ruth's in with the boys. Ollie couldn't sleep."

"I just saw her. She was heading for the beach."

"Don't be daft."

"I wouldn't lie to you."

"Look. She's upstairs. I promise you." He clattered up the stairs and opened a door. There was a pause during which Jess could hear his breath rising, then he slid back down again.

"She's not there."

"No."

"Did you speak to her?"

"She didn't see me."

Ben said nothing. Instead, he stared at Jess, his eyes hooded.

"Has she done this before?"

Ben shrugged. "I'm not sure." He gritted his teeth. "Yes. Yes, she has."

"Don't you think we'd best go after her?"

"What? Oh, yes. Yes, of course."

Jess narrowed her eyes at her brother. What had happened between him and Ruth, that he wasn't immediately hurling himself down to the beach in pursuit of his wife?

"Is everything OK?" she asked.

"It's fine." His voice was terse.

"I'll go, if you want. You stay with your boys."

"No. I need to find her. Can you…?"

"Of course." She smiled. Looking after two five-year-olds was the last thing she felt like right now, but they were her nephews. And they'd be unlikely to wake. She nodded.

Ben reached for the door. "Thanks."

As he opened the door, something fell against it on the other side. He yanked it open.

"Ruth?"

A woman was leaning against the door, her head bowed. She had shoulder-length blonde hair and she was slight. It wasn't Ruth.

"What?" exclaimed Ben.

The woman looked up. It was Sally Angus. One of the women who'd been kidnapped with Ruth.

"Sally," said Jess. "You should be indoors. There's a curfew."

Sally's nostrils were flaring. Her cheeks were flushed, and she was missing a shoe. "You have to come quickly," she panted.

"Can't this wait until the morning?"

"No. You have to come now. And Ruth. Mark needs her."

"Mark?" Mark was Sally's fiancé. An annoying, wheedling man who they'd stopped from coming along on the rescue mission. "Have they come back?" Jess felt a fist clutch at her chest. "Is it happening again?"

Sally shook her head. "No. It's that Martin. He's attacked my Mark."

31

Jess stared at Sally.

"What?"

"You heard me. He attacked him. Mark's face is a state. He'll get infected if Ruth doesn't help him."

Ben's voice was low. "I told you he was trouble."

As Jess recalled, Ben had done no such thing. She gave him a warning look then turned to Sally. "Did you see this attack?"

Sally straightened her shoulders. "Yes. It was when Mark was coming home from the shop. He'd picked up our rations, ready for the curfew. Then he was walking across the grass outside our house and Martin jumped out at him. He covered him in scratches." She wrinkled her nose. "He fights like a woman."

"You're sure it was Martin?"

"Of course I'm bloody sure. What are you accusing me of?"

"I'm not accusing you of anything. But *you* are. You're accusing Martin of a serious offence. If he really attacked another villager..."

"What do you mean, *if he really attacked a villager*? Are you saying I'm lying?"

Jess stared up at the ceiling for a moment, pulling together all of her strength. She looked at Ben. She wanted to urge him to go after Ruth, but she didn't want Sally to know what was going on.

"Sally. There's a curfew on, like you just said. I think it's best if you stay here for now, keep indoors. We can sort this out in the morning."

"I'm not staying here." Sally gave Ben a look of disgust, as if she might catch a communicable disease if she took shelter in his house.

"I'll stay too," said Jess.

"Mark needs me. He needs you to talk to him. To hear what happened. Every minute that Martin stays here is a minute we're all at risk."

Jess stared at Sally. Did she really think like this? Then it occurred to her how little effort she'd taken to talk to Sally after the kidnappings, to find out how she was recovering.

But still.

"It's not as simple as that," she said.

"He attacked my husband."

"I'd need to hear both sides of the story."

"It's not a story."

Jess looked at Ben. He was staring at Sally, his jaw firm. "It'd be easier if we just kicked him out," he said.

"No," said Jess. "He faces the same justice as anyone else. We need to find out the truth, and then we can decide if there's punishment to be meted out."

"Oh, this is ridiculous," snapped Sally. "I'm going to tell him what we think of him."

She turned and started running away from the house.

"Sally, no!" Jess called.

"I have to go after her," she told Ben.

"What about Ruth?"

She let out a long, shaky breath. "I'll see if I can find her, after I've stopped Sally. If Martin's locked his door, there shouldn't be any trouble."

"Be quick."

She squeezed his arm. "I'll be as quick as I can," she said, and started running after Sally.

32

SARAH HAMMERED on the door to Martin's flat. It flew open.

"What is it?" Martin's face shifted from fear to irritation. "Oh. Didn't you have your key?"

"It's not that. It's the people."

"What people?"

She pushed inside and bolted the door behind her. She was hot, despite the drizzle she'd run through. "From Filey."

He paled. "They're back."

She nodded. "I have to go back to Jess."

"What do you mean, go back?"

She cursed herself. "Nothing. I'll go and tell her."

"You're soaked. I'll go."

A shout came from outside.

"Oi!"

Sarah frowned at Martin, her eyes widening. He'd paled.

"Who is it?" she whispered. "Is it them?"

He put his fingers to his lips and crept to the window,

bending over so as not to be seen. He crouched beneath it and lifted his head slowly to see out.

"Who?" she hissed.

He raised a hand behind him. "There's a woman out there."

"Martin! Get out here now!"

"Is it the attackers? Oh God, they saw me."

"They would hardly know my name," said Martin.

She felt her stomach dip. He was right. It had to be the same person who'd left them the note, and the 'present'.

She followed him to the window and looked out, not bothering to hide.

"Sarah! Get down."

"It's only Sally Angus. What's she doing here?'

"Your bloke attacked my husband!"

Sarah frowned down at the other woman. Sally had been on the other side of the wall, in the farmhouse. She had no recollection of a wedding.

"What are you on about?"

"What's all that shouting?" A man's voice this time, from directly beneath them. Keith, who lived in the flat below.

"Your neighbour's a monster!" Sally cried.

"Go away," came the reply. "Get indoors."

Martin was at Sarah's side now, the two of them looking out at Sally.

"Have you seen Mark Palfrey today?" she whispered.

"I don't even know who Mark Palfrey is."

She pointed at Sally. "Do you remember her? She was at the farm."

Martin nodded.

"Mark is her fiancé. Husband, if you believe what she's saying."

"I've never set eyes on the man."

She eyed him. Martin had no reason to lie. If he said he didn't know who Mark Palfrey was, then that was the truth.

"Sally, go home!" she called. "You need to get indoors. Martin doesn't know what you're talking about."

"You can't fob me off like that."

Sally disappeared. Sarah heard thumping on the door downstairs, the outside door to their block. She grimaced, thinking of Keith downstairs. "Why won't she just go away?"

"Maybe if I speak to her, she will," Martin said.

"I don't think that's wise."

"We can't have her out there disturbing the neighbours all night. And what if it kicks off again? She won't be safe."

"Leave it. You have to report to the JP anyway. Mum volunteered you to be one of the guards tonight."

But he hadn't heard her. He'd opened the door to the flat and disappeared down the stairs. Sarah followed.

"Stay upstairs," he said, shooing her away. It's safer up there."

"Don't mollycoddle me."

He said nothing, but let her stay close to him as he opened the outer door. Sally was outside, yelling obscenities. Jess was with her now.

"Jess? I was just heading to the JP. Guard duty." Martin sounded shocked, and not a little embarrassed. Sarah thought of their conversation earlier, of the look on Jess's face when she'd mentioned Zack.

"It's not that," Jess said. "Sally says you attacked Mark."

Martin sighed. "I don't even know who Mark is."

"Liar!" Sally screamed. She launched herself at Martin, grabbing the collar of his shirt. Sarah gasped as it tore. Martin stared down at Sally, but did nothing. Sally was tiny; not much more than five feet tall, and slim. If he retaliated, he would hurt her.

"Right," said Jess. "I don't need this, tonight. Martin, Sarah, I want you back indoors. We'l get someone to cover for you, Martin. Sally, I'll take you home. You can show me any injuries Mark might have, but I'm not going to follow this up until the morning."

"That's ridiculous," said Sally. "By the morning, he'll—"

"I wasn't inviting debate," said Jess. "Now, are you coming with me, or not?"

Sally seemed to shrink into herself. Sarah allowed herself a smile at Jess. Not everyone knew what Jess was capable of, but Sarah knew how strong she was. If it hadn't been for Jess, Sarah would have nothing. No family, no boyfriend, no job.

Jess turned to her. She didn't return Sarah's smile. "You two, get indoors. Now."

Sarah pulled Martin inside. She bolted the door then dragged him up to the flat. They locked the door then leaned a chair against it. After a moment staring at it, Martin pulled the chair away and dragged a chest of drawers over to take its place.

"Happy?" Sarah asked.

"No."

She slipped her arms around him from behind. "It's nothing," she reassured him. "God knows what Sally's got into her head, but Jess will see that Mark's got no injuries, and it'll all be over in the morning."

"I hope so." He was looking at the door, no doubt wondering how someone had got inside the block to post anything through their door in the first place. "Thanks for not telling her about the notes, and the shit."

She nodded, feeling her chest tighten. She hated lying to him. She'd spotted Jess looking at the hole in the window. But true to her word, she'd said nothing.

Then she remembered. "Oh my God."

"What?"

"I didn't tell her. About the people from Filey. I have to go out."

He grabbed her hand. "You can't go out there."

"What if they attack again?"

"What did you see?"

She shook her head. She didn't know everyone in this village; she'd spent too much of her life shut indoors. But they'd been smoking. And they'd talked about not being seen.

"They were in one of the houses. By the Meadows. It looked – I'm not sure – it looked suspicious."

He tightened his grip. "What was suspicious?"

"They were standing, smoking, outside one of the houses. That's why I noticed them. No one smokes here."

"No. What else?"

"There were people inside. More than Jess said. Six of them, I think. Maybe more. Plotting something, it looked like. They talked about not being seen. It felt wrong." She turned to him. "It could be nothing."

"Doesn't sound like nothing." His grip slackened. "Wait here. I'll go."

33

RUTH SAT on the damp sand, her limbs numb.

She'd checked on the boys before leaving and she knew they were asleep. So was Ben.

Lying in bed next to him, all she could hear was the sound of Robert's breath, heavy in the confines of his neat bedroom at the farm. Every time she looked at her husband, she saw him. The way he stared into her face like he was trying to worm his way into her brain. The way he had stroked her skin, making her want to climb out of it.

This afternoon, Ben had come into the pharmacy. He'd seen how pale and tired she looked and tried to give her a hug. She'd felt like she would suffocate. She pushed him off, moaning, and then been pricked by guilt when she saw the hurt on his face. How could she tell him that every time he came near her, it wasn't him there but his childhood friend? And that it was all his fault?

The only people she could bear to touch were her sons. She'd approached them hesitantly this morning, after her reaction to Zack's death, believing herself to be a monster. She'd killed a man. She felt angry and confused every time someone showed her affection. Could she hurt her children too?

But to her great relief, holding them in her arms had quietened the rage that churned inside her, not ignited it as she'd feared. But when they spoke to her, when they asked her questions or looked to her for reassurance, she knew she was found wanting. She'd lost the ability to gauge their reactions to anything she said, the skill she'd had for reading between the lines of their smooth faces and understanding what they weren't asking from her, but needed nonetheless.

She couldn't tell anyone how she felt. Jess was too full of grief over Zack. Ben was the enemy now, someone she had to avoid, and keep at a distance. And everyone else needed her. Six people had been injured in last night's attack. Six people who'd had to wait all night and much of the morning before she could bring herself to treat them. Treating minor wounds without making skin contact wasn't easy. She'd worn a pair of woollen gloves from home, making excuses about having an allergic reaction to something or other, and hoped the fibres wouldn't get into any of the cuts she was treating.

The village didn't know that Ruth had gone. They didn't know that she'd been replaced by this woman who wanted to kill everyone who came near her. Who wanted to blank it all out.

She stood up and waded into the shallows, ignorant of the cold. Ruth had grown up near the coast, and had always had a healthy respect for the sea. She wasn't a strong swimmer. If she went out there, she wouldn't come back.

Would her boys survive without her? Would they be better off without a mother who was incapable of feeling warmth?

She was up to her knees in the dark water now. It lapped around her legs, splashing at her as if trying to get inside her skin. It felt light and heavy at the same time, cold and oddly warm. She bent to run her fingers through it, shuddering at the movement against her hand. She took a step forwards.

"Ruth?"

She froze, her eyes piercing the darkness.

"Ruth!"

The voice grew louder. She heard stumbling behind her, the sound of someone running across the damp sand.

It was Ben.

"Ruth, what are you doing?"

She took a deep breath. She couldn't turn around. She couldn't see that look on his face. He would be looking at her like this was her fault. Like she was the one in the wrong, the one whose mind was lost.

When would he take responsibility for what he'd done to her?

He was at the edge of the water now, just feet away. She could hear his clumsy movements in the shallows.

She raised a hand to stop him.

"Ruth, I don't know what you're doing. Jess said she'd find you. But I couldn't wait."

Wait for what? For her to walk into the sea like it would swallow up her rage and bewilderment?

"Ruth. Please, come home. It's not safe. The boys need you."

There was panic in his voice, and could she hear guilt?

"I need you."

There it was. His need, bigger than her own. Looming in his mind like there was nothing else that mattered.

"You never ask what I need," she muttered.

"What? Did you say something, love?"

She felt her face crease at the 'love'. *Don't call me that. You have no idea what love is.*

His hand brushed against her arm. She yanked it away.

"Ruth love, you're scaring me. Come home, please."

The cold was seeping up her body now. It gnawed at her organs, and trickled into her arms. She shivered violently. She felt something land on her shoulders. A coat. She went to throw it off then stopped herself. It was just a coat. Ben's coat. But still, it couldn't hurt her.

She turned to him, her eyes lowered. "Don't touch me."

"What's wrong? What did I do?"

She said nothing.

He advanced and she shook her head, pulling her neck back. "No."

He nodded. He looked like he might cry. She thought of their sons, their tousled heads on their pillows as she'd left the house. She wiped a tear from her cheek.

"Alright," she said. She lifted her feet high to walk through the water, feeling the damp sand suck at her shoes.

She didn't look at Ben as she passed him. He reached towards her but then thought better of it and pulled his hand back. It was shaking.

She could hear dim sounds ahead. Footsteps. Voices?

Had Jess come to find her after all?

"Alright," she repeated. She would go home. To her boys.

She made for the coastal path towards the house, not caring if he was behind her or not. As long as he didn't touch her.

34

Mark Palfrey sat on a hard-backed chair, rubbing a red mark on his skin. For all Jess knew, it was caused by the rubbing, not by any alleged attack.

He sat side-on to Jess and refused to meet her eye. Sally hovered behind him, her fingertips fluttering over the fabric of his T shirt.

"So what happened then?" asked Jess. She considered whether she should have brought another member of the council with her. No, that was ridiculous. This was a minor alleged assault, not the Old Bailey.

"That Martin attacked him," said Sally. "He's bad news."

Jess fought down the irritation. "I was asking Mark."

Sally gave her a hard look. "Tell her, Mark." Her voice was clipped, her West Country accent stronger than usual. Jess realised she knew nothing about this couple's history. Had they been together before they came here, or had they met in the village? And did it matter?

"Mark?" Jess said. "I don't have long, so I'd be grateful if you could fill me in."

He sniffed. “You need to hold a village meeting.”

“I don’t see what that’s got to do with anything.”

He turned to her. There was a red line snaking over his right eye. It looked swollen.

“Have you been to see Ruth for that?” she asked.

“Didn’t you hear me? You need to show everyone what he’s like. You let him in.”

She placed a hand on the back of the chair next to Mark's. She was standing over him, having not been invited to sit. “You still haven’t told me what happened.”

“I think it looks pretty obvious, don’t you?”

“Martin attacked him,” Sally interrupted.

“You’ve already said that. But I need more details. When? What was the weapon? And why?”

“Why? Because he’s no good,” said Sally. Jess narrowly avoided rolling her eyes.

“You have to understand. You’ve clearly been hurt, Mark, but before I can accuse another member of our community”—she ignored Sally’s huff — “I need more information. I need evidence. You say Martin did this to you. Can you tell me what he used, and when it happened?”

“His nails, of course,” said Sally.

“Mark?”

Mark nodded. Jess bent over him to look more closely at the wounds. They did indeed look thin and ragged, like they’d been created with fingernails. She glanced at Mark’s fingers; his nails were neatly trimmed, almost down to the quick. Sally’s were longer, and a couple of them looked as if they'd been bitten.

“OK,” she said. “I’ve got other things to attend to. You come and find me when you’ve got more to say. OK?”

Mark said nothing. Was this the same man who’d found it impossible to shut up when his girlfriend had been kidnapped? Was it the effect of the alleged attack, or something else, that was making him taciturn?

"OK?" she repeated.

"Don't worry," said Sally. "We'll find you."

"Good."

She let herself out and made for the centre of the village. She passed Martin's flat and looked up, wondering if Sarah and Martin were up there, looking at her. But the window was dark.

Ahead of her, two figures were moving up the hill form the beach. She stiffened; another raid? But then the one at the front spoke: it was Ben, with Ruth. He'd found her. She allowed herself a relieved smile and decided to leave them to it.

She sighed. It felt more like being back in the schoolroom than running a community of adults, with people accusing each other of things and launching petty attacks. Maybe she should sit them all in separate corners for a day until they learned to behave themselves.

She approached Toni's flat. The curtains were drawn. She crept upstairs, anxious not to disturb Toni and Roisin.

Btu the flat was dark, and quiet. She shuffled around the kitchen and poured herself a glass of water.

Toni emerged from her bedroom, yawning.

"Hey."

"Hey," replied Jess. "Everything OK with Roisin?"

"Yeah." Toni rubbed the back of her hand across her eyes. "Just don't tell anyone she was here, huh?"

"None of my business."

"Thanks. You can have the bedroom, if you want."

"It's fine."

"You sure?"

Jess's head was full: of Mark and Sally, of Ben and

Ruth. Of Zack. Sleeping on a sofa would be easier; a bed would feel so empty.

"Sure. See you in the morning."

Toni slipped back into the bedroom. Within moments, Jess could hear snoring. She found the pile of blankets Toni had left out for her and arranged them on the sofa. She settled down, fidgeting.

She heard rattling coming from the window. The wind was picking up; she had no idea Toni's windows were so loose. She'd ask the maintenance team to fix them tomorrow.

There it was again. Rattling. She pushed the blankets to the floor and opened the curtains.

She flinched as a hail of pebbles hit the glass in front of her.

"What the hell?"

She opened the window. "Sally? I told you I'd deal with it in the morning."

"It's not Sally." Martin's voice.

She sighed. "You want to know what happened with Sally and Mark."

"No. I came to warn you."

"Warn me about what?" Not more people posting things through letterboxes, she hoped.

"It's the people from Filey. We think they're starting something."

This was all she needed. "I'll be right down."

They hurried towards the road where the newcomers had been housed. Drizzle still filled the air and the wind gusted off the sea. Jess longed for her bed, even if that bed was just Toni's sofa.

They rounded the road where the newcomers had been billeted and stopped to look at the houses. They blinked back, impassive. She wondered if the occupants were watching.

"Are you sure?" she whispered.

"Sarah said she saw people. More than just the people who should be here."

"How many?"

"She wasn't sure."

This was ridiculous. "There's nothing here. They're asleep."

She grabbed his arm and guided him away from the houses. "Tell me what she saw."

"There were people standing outside one of the houses smoking. And more people inside."

"They could have been visiting each other."

Martin shrugged. The sense of purpose seemed to have left him.

"I can't wake them up in the middle of the night without a good reason," she said. "I'll come back in the morning."

She started walking. Martin stayed where he was, staring at the houses. She ignored him and hurried to Toni's. She had no idea who to believe anymore.

35

THE MORNING WAS dull and cold, a biting wind scudding in from the north. Jess headed for the newcomers' houses, feeling uneasy.

She would knock on their door, pretend to be checking they were settling in.

She rounded the JP. Her legs felt heavy and her mind numb. Being in charge of a village full of people she'd known for years was bad enough, but now there were strangers to contend with. Would they trust her, or would they be as wary of her as she was of them?

As the first house came into view, she stopped walking. Its dark grey-painted facade was daubed in graffiti.

She looked along the road – no one around – and hurried over to the house. Across the front wall, in large, ragged letters, were the words *Go Home*.

She put a hand on the *e* of *Home*. It was dry, painted in a bright white that gleamed against the grey. This wasn't fresh. But it hadn't been here last night.

She thought of Martin, standing behind her, staring up at the houses. Surely not?

She looked along the row of houses again. Would anyone have seen?

She scratched at a patch of dry skin on the back of her hand. Goosebumps pricked her arms; the cold was deepening.

Should she knock on the door, let the inhabitants know what had happened? She stepped back and stared at it, trying to sense whether people were moving around inside. The two windows facing the road had their curtains closed.

She looked along the row of houses, towards the flats. Surely Martin wouldn't have done this? He was angry, and scared. But graffiti?

The people inside should be told. They probably needed to be protected. She heaved a deep breath at the irony. She had no idea who would help with that.

But the village had one person who could be relied upon to follow the rule of law, or at least of the village rulebook, and push his feelings down. That person was Colin.

She strode to the Barkers' house and rapped on the door three times. Sheila opened the door, her eyes wide.

"Oh, Jess, it's you. I thought you were Ruth."

"Is Colin in?"

"He's just eating his bacon. Come in."

Jess smiled and let Sheila usher her in.

"I'm heading to the school. I'll leave you two to it." Sheila tugged on a bright pink hat and stepped out into the morning.

"Sheila."

Sheila turned. "Yes?"

"Which way do you walk, to the school?"

A frown. Sheila pointed in the direction of the coastal path. "I take the long route. Give Benji some exercise."

"Good." That meant she wouldn't pass the grafittied

house. Even so, this needed to be dealt with quickly. People would be starting the day soon. "I mean, good for you." Jess gave Sheila a tight smile. "See you later."

Sheila shrugged and trudged off towards the beach, humming under her breath. Jess turned back towards Colin. He was sitting at the kitchen table, cutting a slice of bacon into neat squares.

"You're not going to like this."

He forked two squares into his mouth and swallowed. "I thought as much. Bit early for a social call."

"Sorry."

"Did we miss another attack?"

"No. It's the house. The one we put the people from Filey in."

"The Snellings want it back?"

"No."

"What, then?"

"Finish your breakfast, then come and look."

Colin grabbed a slice of heavy brown bread and put the bacon squares into it. He palmed the sandwich he'd created. "Come on. I can tell you're in a hurry."

When they reached the house, he whistled. "Oh, no."

"Oh, no. Is that all?"

"I was expecting worse."

"Like what?"

"A fight, or something. I guess."

"What are we going to do?"

"Clean it off would make sense."

"That's not what I mean."

He stared at the letters. "I don't know, Jess. I really don't. This is no worse than what their lot have been giving us for the last six years. They should be glad it wasn't worse."

"That doesn't make it right."

The front door of the house opened a crack. A man looked out, his movement wary. When he caught sight of Jess and Colin, he stiffened.

"Who are you?"

"I'm Jess Dyer. Village steward. I met your wife yesterday. This is Colin Barker, secretary of our council."

"What are you doing here?"

"I'm afraid your house has been vandalised."

"It's not my house."

"I'm sorry. But you need to see."

He stepped outside and stood next to her. He was wearing a blue sweater that looked home-knitted, and heavy brown cords. His eyes roamed over the words on the wall of the house he'd slept in.

"Bastards," he said. "I'll fucking kill them."

"It won't come to that," said Jess. "But we will find out who did it, and in the meantime we'll help clean it up."

"Like hell you will."

"Sorry?"

"This is evidence."

He ducked back into the house and emerged holding a mobile phone. He started taking photos.

Colin tried to take it off him. "Mr Haywood. I really don't think this is a—"

"Get your filthy hands off me! That's assault."

"Please. We can help with this."

"Colin," Jess muttered. He withdrew his hand and stopped trying to take the phone. Better to let the man take his photographs than to risk a fight.

When he'd taken his fill of photos, the man turned to them. "Now leave us alone."

He retreated inside and slammed the door.

Jess knocked on it. There was no response.

"Let's just leave it," said Colin. "They can clean it up themselves."

"You're kidding. What will happen when everyone sees this?"

"Why do you care?"

"Because however obnoxious that man is, he's part of this village now. We don't want things getting any worse."

Colin sniffed. "Let's try the other house. You said they were friendly."

Jess knocked on the door of the Bagris. She hopped from foot to foot, wishing she'd put on a warmer coat.

Once again, there was no response. She stood back to look up at the upstairs window. The curtains were open. The downstairs window was open and there was no sign of life inside.

"This makes no sense," she said.

"Maybe they're asleep."

"No. This smells wrong."

"How?"

"Sarah saw suspicious activity here last night, and now the Bagris have gone, by the looks of it."

"We don't know that."

"You need to fetch the master key," she said. "We have to get inside."

36

"I DON'T LIKE THAT IDEA," said Colin.

"I knew you'd say that," replied Jess, "but we need to know if they're OK."

"Why wouldn't they be?"

"I spoke to Mrs Bagri earlier. Aabida. And her husband. She seemed scared. Said the people next door hated them."

"That doesn't give us a reason to break into their house."

"We've knocked three times. They're either gone, or they aren't coming out for some reason."

"Isn't that their business?"

"Sarah Evans said she saw suspicious activity here last night. I'm worried."

"What kind of suspicious activity?"

"People. More of them than there should be."

"OK."

Jess heard a voice from the other house, the Haywoods'. She put a hand up to listen, but there was nothing more.

"Colin, please. I'll take responsibility. I just want to go in there and check that everything's OK."

Colin looked at the Bagris' front door again and sighed. "If there's any trouble, or anybody asks, you coerced me."

"That's fine."

Ten minutes later, Colin had returned with the key. Jess watched the houses while she waited for him, backing away to stand on the opposite side of the road. The Haywoods' house was silent, the curtains unmoving. The Bagris' was just as quiet but with the curtains open. It felt wrong.

"Here." Colin handed her the key. "You can do it."

"Right." She unlocked the Bagris' front door, glad of the original cleaners' master key.

She pushed the door open a short way.

"Hello? Mr and Mrs Bagri?"

No answer. She stepped inside.

The hallway was empty. No sign of the toys she'd seen behind those children yesterday. The hooks were empty of coats and there were no shoes on the floor, as there always were in Ben and Ruth's house.

She went into the living area. "Hello?"

It too was empty. The tired sofa that sat in the middle of the space looked as if it hadn't been sat on for months, even though another couple had been here until very recently.

The kitchen too was bare. No food on the worktops. No sign of the rations they had provided for the Bagris.

"Where the hell are they?" asked Colin.

"Shh." She raised a finger to her lips. She could hear voices through the wall. Men. She couldn't make out what they were saying but there was definitely more than one man in there. Even if she accounted for the teenage sons, there were too many voices.

Then she heard a baby crying, followed by a woman

shouting for it to shut up.

Colin's eyes were wide. "What's going on in there?"

He hurried out of the house and started pounding on the front door of its neighbour.

"Let us in!"

Jess stood behind him. "Calm down, Colin."

He turned to her, a stray lock of hair bobbing over his eyes. "We allocated that house to one family. They can't start moving more people in. It's against the rules."

"I think we're past the rules now."

"What do you think they're doing in there?"

She thought back to the night of the attack. She hadn't seen the faces of any of the invaders; she had no way of knowing if the Haywoods were among them.

Surely Anita Chopra wouldn't have brought people here who'd attacked them?

But maybe Anita didn't know. Or didn't care.

Colin pounded on the door again.

"Open it," Jess said.

"It's occupied. I can't."

"Colin, please. Forget the rules for once."

He pulled the key out of his pocket and handed it to her. Then he pushed her hand down.

"No." He held onto her hand. "I have to believe that someone is in danger before I let you break in."

"I already said we're not breaking in. We've got a key."

"You know what I mean. If you do this, I'll have to report you. You'll be kicked off the council."

She dragged a hand through her hair. Why did Colin have to be like this?

"Can I overrule you?" she said. She pocketed the key, frustrated.

"You'll have to call an emergency meeting of the council."

37

The room that had once been a coffee shop but now did double duty as school by day and village hall by night was dull in the early morning light. Jess watched as the council members filed in. Normal conversation that accompanied the beginning of council meetings was absent. Today, people were withdrawn and slow-moving. Faces were pale and eyes were dark-rimmed or bloodshot. Ben, opposite Jess, stared at his hands. He twisted them together in front of him, muttering. They were all standing, or perched on the edges of tables; there hadn't been time to rearrange the furniture.

"You OK, Ben?" she whispered.

He looked up. "Let's get this over with, huh?"

She nodded. "How is she?"

He shrugged. "I don't know, sis. She won't talk to me. Won't let me anywhere near her."

"She's been through a lot."

"We all have."

"Would it help if I spoke to her?"

He shook his head. He opened his mouth to speak but stopped as Colin called the meeting to order.

"Thanks for coming at short notice, everyone. We've got a problem."

Heads turned towards him and the silence thickened. Jess wiped her damp palms on her jeans.

"Jess?" Colin prompted.

She nodded. The council members stared at her, waiting.

"The people who've been housed in the cottages over by the Meadows," she said. "The ones from Filey."

"We know who they are," said Ben.

She glared at him. "I think there's more of them in there."

Ben tensed. He stared at her, his cheeks red. He would be angry that she hadn't told him first.

"What's happened?" asked Dawn.

Jess stood up. She swallowed. "An hour ago, I spotted something on one of the houses. Graffiti." She surveyed the room, wondering if anyone here knew who had done it. Or if it mattered. "But they wouldn't let me and Colin in. The Haywoods. I think there are more people in there. And the other house, the Bagris. It's empty."

"Empty?" asked Sanjeev.

"I knocked on the door and no one answered. We were worried, so Colin got a key."

She glanced at Colin. He was holding himself very still.

"There was no one in there," she said. "But we could hear the other house through the wall. There seemed to be a lot of people inside."

"Maybe the Bagris had gone next door to be with the Haywoods?" suggested Dawn.

"They hated each other. I spoke to Mrs Bagri. She was scared."

She leaned on the table behind her. "I have no idea what 's happened to them."

"What about all these other people?" asked Ben. His eyes were bloodshot and he was wearing the same shirt she'd seen him in last night.

"I don't know," she sighed. "All the curtains were closed and the deadbolt was closed. You know the rules about letting ourselves into an occupied house."

Colin sniffed.

"We have to go there," said Ben. "Right now."

"Let's not be hasty," said Jess. She surveyed the group. "Where's Toni?"

Sanjeev looked towards the empty seat where Toni normally sat. "Working, maybe."

Toni worked on the allotments, bakery and smoke house, overseeing the production of food. Jess frowned; it would be good to have the support of her friend today. But she couldn't afford to wait.

"Who's to know they aren't plotting something?" said Harry. Harry was one of the older members of the council. He'd been an ally of Ted Evans, Dawn's husband. He'd mainly kept himself to himself since Ted's arrest.

"Why would they be doing that?" asked Jess.

"The graffiti could have been a diversion," said Ben. "Keep us busy while they tell their friends from Filey about the layout of the village. Our weak points."

She frowned at her brother. "You're being paranoid."

He shrugged. "I'm being careful."

"The reason I brought this up wasn't to discuss your conspiracy theories. It was to ask for the council's permission to go into that house. And to start the process of identifying whoever did the graffiti. I don't care who the victims were; it's not acceptable for anyone to deface another villager's house."

"They aren't villagers," said Ben.

She said nothing.

"I'll see what I can find out," said Dawn. "They're neighbours of mine. I've spoken to Mary once or twice."

"We don't have time for that," said Ben.

"Ben, please—" began Jess.

He stood up. "No. If they've brought more people, who knows what they're up to. Your friend Anita Chopra said nothing about more people."

"She isn't my friend."

"Whatever. Come on. I'm going over there now."

Jess approached him. She lowered her voice. "Is Ruth at home?"

He paled. "Yes."

"Don't you think you should be with her, instead of careering around the village? I'm worried about her."

"You think I'm not?"

"That's not what I said."

Colin cleared his throat.

"Sorry," said Jess.

"I'm going over there," said Ben.

"Not just yet," said Jess. "There's a second agenda item. An alleged attack."

"The people from Filey?" asked Dawn.

Jess shook her head. "Mark Palfrey was assaulted, yesterday, he claims. His girlfriend, Sally Angus, says that she knows who did it. But so far there's no evidence. And Mark himself won't say anything about it."

"Who did it?" asked Sanjeev.

"I can't be sure."

"Who does Sally say did it then?" asked Colin.

She threw him a glance. She should have talked to Colin about this beforehand, but they'd been too caught up in the graffiti incident.

"Martin Walker," she said.

"What?" Ben was standing by the door. He looked as if he might laugh.

"She *claims* it was him. He denies it, and as I say, Mark hasn't said anything himself."

"Well it's hardly surprising, is it?" replied Ben.

"Ben," warned Colin.

"No." Ben stood up. "Look, I was taken in by him last time. I let him sleep in my house. You all know how that turned out. We should never have let him back here."

Dawn moved away from the chair she'd been leaning against. She looked from Jess to Ben. Her eyes were bright and her chest rose and fell conspicuously.

"Dawn?" said Jess. If Martin did attack Mark, she realised, then Sarah might not be safe. "I should have told you about this.".

Dawn shook her head. "I don't believe them."

"Sorry?"

Dawn stood. "When Martin came here, I was scared. I knew what he did." She looked at Ben. "To your lovely wife. I'm not denying that. But he was coerced. He's not like that. I've seen him with my daughter. He's kind, and tender. He wants to stay here, and to prove himself."

"He's not doing a very good job of that," said Ben.

"Did you see him when we were attacked?" asked Dawn, her cheeks pink. "He fought those people off as bravely as anyone who's been here all along. He's one of us. I know it."

Jess bit her lower lip. It was swollen, and tasted metallic. If Martin hadn't attacked Mark, then who had?

"Did anyone see Mark during the attacks?" she asked.

Heads shook all round.

"We need to find out if he was hurt then. Ben, do you know if Ruth treated him yesterday morning?"

Ben shook his head. "Mark Palfrey is one of us. Don't we trust our own anymore?"

"Martin is one of us, too. If one villager accuses another of attacking them, we owe it to both of them to find out the truth. We can't take sides."

"Bloody naive."

Colin coughed. "Now, Ben. This is a council meeting."

"I don't care if it's the palace of bleeding Versailles. Martin took my Ruth, and now he's assaulting our people. He's no better than those murderous bastards from Filey."

This was going nowhere. "I'll talk to them all again," Jess said. "Mark and Sally, Martin and Sarah." She noticed Dawn shrinking into herself. "If you want to be with your daughter, Dawn, that's fine."

"Thank you."

"I'll get to the bottom of it."

"We all know what you'll find," said Ben. His eyes were small and dark and his shoulders hunched.

"Let's not make assumptions," she replied.

"I'll come with you," said Colin.

"Sorry?"

"That's a good idea," said Sanjeev. He was watching Ben. "We don't want Jess being accused of bias."

I'm not biased, she wanted to snap. But Sanjeev was right. An impartial observer like Colin would protect her from any accusations.

"Fine," she said. "Colin, I'll see you back here in an hour. But we need to go to those cottages first." She tried to face the other council members. "Do Colin and I have your permission to use the master key, if they still won't let us in?"

Mutters of assent. As she turned for the door, it flew open. She shrank back.

Toni stood in the doorway, out of breath. “Sorry I'm late,” she said. “Did I miss anything?”

Jess stared at her. “You could say that.”

Toni was flushed and her coat was in disarray. She blinked at Jess a few times, as if not entirely present.

“Get back here!”

Toni stumbled into the room as someone barrelled through the door behind her. The council members shrank back.

Jess stepped forward. One of the newcomers? But no, it was Flo Murray. Roisin's mum.

Flo barged into the room and looked quickly around. Her gaze landed on Toni.

“What in heaven do you think you've been playing at?” she shouted.

38

Toni barged past Flo and into the square. The council members followed, suddenly quiet.

"Flo, please," Toni said. "This isn't the time."

"I don't care if it's the time or not," Flo replied. "My daughter has been through enough. And now she's lying to me, because of you."

Jess took a step forward. "Flo, I'm sure this isn't as bad as you—"

Flo turned on her, her eyes wild. "You've been helping them, haven't you? And you the steward too."

"I don't know what you're talking about."

Jess looked at Toni. "Where's Roisin?"

"At home. She's been grounded. Like a kid," she muttered.

Jess nodded. "Please, Flo. We're in the middle of a village meeting right now. Can this wait?"

Flo ignored her and advanced on Toni.

"Leave her alone, you hear me? She's still recovering from what those awful men did to her. Do you have any idea?"

Jess saw a shadow pass over Toni's face. She probably knew better than anyone what Roisin had been through.

"She needs peace and quiet," Flo continued. "Not—this!" She waved a hand in dismissal of Toni.

"Is this because I'm a woman?" accused Toni.

"Woman, man, whatever! She doesn't need you. She told me she wasn't going to see you anymore, and now I find out she's been sneaking about behind my back."

"She's a grown woman. She's allowed to—"

Flo jabbed a finger into Toni's face. "She's twenty. She still lives in my house. Don't you go filling her head with nonsense."

Jess pulled Toni away from Flo. "Maybe if I can sit down with you all later, and we can work something out?"

"There's nothing to work out," spat Flo. "You keep that woman away from my girl, or I'll tell people the truth about your family."

Jess felt the air behind her move. The entire village council was standing in the square now, seemingly oblivious to the cold. Someone muttered.

"What truth?" she asked.

"I know why he did it. That Robert Cope."

Jess narrowed her eyes at the other woman. For herself, she didn't care who knew about Ben and Robert, and their past. She didn't care on Ben's behalf; he'd made that bed years ago. But Ruth…

She approached Flo. "Please, for Ruth's sake. Let's just calm things down a bit, eh? I'm sure we can find a solution."

"Help!"

Jess span round at the voice. Sheila stood behind them, her cheeks pale and her breathing laboured. Colin stepped towards his wife, looking perturbed. Flo stared at Sheila, angry that her thunder had been stolen.

"Sheila? What's wrong?" Jess asked. "Is it the school?"

Sheila shook her head. Her normally tidy if coarse hair was haloed around her head, curls sticking up in all directions.

"Where are the children?"

"On their way back. Sarah's with them."

"Way back from where?"

Sheila looked at Ben. "Oh my God."

Colin strode to his wife. "Sheila love, tell us. What is it?"

Sheila swallowed. She looked from Colin to Ben, and then back to Jess. "It's Ruth. We just found her, on the beach."

"What's she doing there?"

Sheila scanned the room again. She opened her mouth but nothing came out.

Ben stood up. "Sheila, tell us." He pushed past her, almost knocking Sanjeev over in his haste.

"We found her, on the beach," Sheila echoed. "Little Paul Murchison, he was paddling." She threw her hands to her face.

Jess tried to keep her voice steady. "Sheila, please. What do you mean?"

"Ruth. She was in the water." She dragged her fingers down her pale skin, pulling the flesh out of shape. Colin put an arm around her. She ignored it. "I— I think she's dead."

39

MARTIN WAS in the flat when Sarah came home from school.

"Back early?" she asked, trying to inject levity into her voice.

"Akash sent me home." Akash was Martin's supervisor on the allotment.

"Why?"

"Said there wasn't enough work to do."

"That makes no sense."

He was in the kitchen, making tea. He poured her a mug and placed it in front of her. "He didn't want me there. Someone's got to him."

"You think so?"

She hugged him; he smelt of sweat and dirt. Normally he changed out of the clothes he used for work as soon as he came home.

"But you're going back tomorrow, right?" She sipped at her tea, realising how thirsty she was after a day using her voice.

He shrugged. "He said he'd let me know in the morning."

"What?"

He stroked her cheek. "I have to stop this," he said."

"How?"

"I don't know. But I'm one of you now. I can't be here and not work."

"I'll talk to Mum."

"No. You can't go running to your mum every time I have a problem. I'll sort this out myself."

She heard a slam in the corridor outside the flat. They both stiffened and watched the door to the flat. Sarah could feel Martin's heartbeat against her chest.

Silence. They let themselves breathe again.

Martin slid towards the door and pulled it open a little.

"No one out here. Probably just Gloria getting home." Gloria lived in the other first floor flat.

Sarah nodded; should she knock on Gloria's door, to check?

No. She was being paranoid.

"We can't carry on like this," Martin said. "I'm going out."

"Where? You're going to confront them?"

"No. Something else."

"What?"

"It might not work."

"Martin, tell me what's going on."

He kissed her forehead. "It'll be fine. I won't be long."

40

JESS RAN after Ben towards the beach. Sam Golder was staggering up the hill, carrying Ruth. She was pale, the colour of a dead fish. Jess stifled a cry.

Ben ran forward. "Ruth! Oh my God, is she—?"

"She's breathing," said Sam. "I don't want to drop her."

Ben grabbed her feet and struggled with Sam to the village hall. Jess ran with them, pushing open the door to let them through.

They laid Ruth on a table. Her mouth was partly open and her eyes were closed.

Jess looked around the other council members. Ruth was the only thing close to a doctor here. Who would treat her?

"Does anyone know what to do?" she asked, feeling helpless.

Colin pushed past her. "She's cold. Best get her wrapped up. And we should put her in a warm bath."

Jess nodded. Ben hauled Ruth up from the table and

struggled with her to his house. Jess ran ahead to open the front door. Inside, Ben stared at the stairs.

"I can't carry her up there."

"I'll help."

Together they manoeuvred Ruth up the stairs. Ben had her under the shoulders and Jess under the knees. Colin ran in after them and flung blankets over her. Jess stared at Ruth, waiting for her to stir, to come to life.

At the top of the stairs, they laid her on the floor, gently.

"The boys," said Jess. "Where are they?"

Ben looked back at her, his eyes wide. "I don't know."

"They're down here!" called Colin. "Playing cards." His voice disappeared as he retreated into the living room to speak to the boys. How long had they been alone?

Ben was in the bathroom, running the taps.

"When did she go out?" Jess asked.

Ben said nothing. His mouth was tight and he was looking between the bath and his wife. He kept dipping his hands into the water, testing it.

"How warm should it be?" he asked.

Jess tried to remember back to when Martin had fallen in the sea, when Ruth had nursed him. "Not too hot," she said. "It'll shock her, or something. Just warm."

Ben nodded. He turned on the cold tap. Jess wondered when this bath had last been used: baths were banned, due to the shortage of hot water. She turned back to Jess and pulled the blanket tighter around her. She rubbed her through the rough fabric, willing her to be warm.

At last the bath was half-full. Ben looked at Ruth, his eyes empty.

"We can do this," she said. "Same way we got her up here."

They lifted her again, Ben at her head and Jess her

feet. They shuffled into the bathroom, careful not to knock Ruth into the doorframe or the walls. Gently, they lowered her into the bath. Ben held her shoulders as she slipped in, holding her head upright. He stared into her face.

"Ruth," he whispered. "Ruth, sweetheart. Come back to me."

Ruth gasped, making Ben almost tumble backwards. Jess held her breath. She counted inside her head, not sure why she felt the need.

Ruth's eyes opened. She stared at Ben, her eyes full of fear. Ben blinked back at her. He forced a smile. "It's alright, love. I'm here. Jess is with me."

Ruth said nothing. She blinked at Ben, ignoring Jess. Then she closed her eyes again.

Ben turned to Jess. "I don't know what to do."

"Keep the water warm. Talk to her."

Ben nodded.

"Ruth, it's me, Ben. I'm here with you, in our bathroom. The boys are downstairs, with Colin. Everyone's safe. What happened, Ruth? What happened to you?"

"Don't," said Jess. "You'll agitate her."

"What happened, Jess? How did Ruth end up in the sea? She never goes to the beach."

"She does," corrected Jess. "She has been, recently. You said so yourself."

"Are you suggesting she did this? That she tried to..." Ben sobbed.

Jess put a hand on his arm, anxious that he not let go of Ruth. "I'm sure there's a perfectly reasonable explanation. She couldn't swim, could she?"

He shook his head, tears dripping onto Ruth's arm. "She grew up in Suffolk. I don't know why she never learned."

"We just need to look after her now. Don't think about all that. Keep her warm. I'll go and get help."

"Who? Ruth's our doctor. She's the only person who knows what to do."

"I'll go to Filey."

His face fell. "No."

"Yes. They can help us."

"They can't."

"I have to try."

"I forbid it."

She shrank back, knowing that he couldn't follow her. "I have to try."

She hurried to the landing. Below, she could hear Colin trying to chat to the boys. His voice was stilted.

"They'll help her, Ben," she said. "I know they will."

"No!" he yelled, stuck in the bathroom.

She ran down the stairs.

"Colin."

Colin looked up from his huddle with Sean and Ollie. He muttered something to them, then hurried to Jess.

"How is she?" His voice was low.

"She opened her eyes. She's getting warmer. But we don't know what to do."

"Can I help?"

"I'm going to Filey."

"Is that wise?"

"What else do you suggest?"

"I don't know." He looked back at the boys. Sean was pulling Ollie's hair. "Mum!" Ollie shrieked.

"Go," said Colin. "Get help."

41

SARAH RAN to Dawn's cottage. She knocked on the door, shifting from foot to foot. No answer.

She pulled out her key and let herself in. The house was quiet. Dawn's knick-knacks stared at her from the shelves and countertops. Sarah liked seeing Dawn's personality on display like this. It was like peeling away a layer of skin and finding a new woman underneath. A woman who could show warmth, and wasn't constantly scared.

"Mum?" she called. But it was useless. Dawn was out.

She ran towards the village hall. Dawn would be there, she'd said something about a council meeting.

As she reached the square, Dawn was walking towards her, her head down.

"Mum!"

Dawn looked up and smiled. "Hello, dear."

"I need your help."

A shadow crossed over Dawn's face. "Yes, dear."

"It's Martin."

"What about him?" A pause, during which the colour fell from Dawn's face. "He hasn't hurt you, has he?"

"No. They sent him home from the allotment."

Dawn folded her arms across her chest. "Why do you need my help with that?"

"He got a letter."

"No one gets letters here. No post. Love, I need to get back to Ruth."

"Why?"

"You don't know?"

Sarah shook her head.

"They found her on the beach."

Sarah felt her stomach drop. She teared up. Dawn reached out and pulled her daughter to her chest.

"She's alive. But she's in a bad way. Hypothermia, by the looks of it. They need help with her little boys."

Sarah nodded. "I can help. They know me."

"Sheila's there. I don't think they need any more people, right now. But can this letter of Martin's wait?"

"Someone threw a brick through his window. There was a death threat wrapped round it."

"Oh Lord." Dawn crossed herself. "Is he hurt?"

"No. It happened a few days ago. During the riot."

"Oh. That's good." She frowned. "Sorry, love. It's not good. But it's not urgent. I need to help Ruth now. We'll talk about this later, yes? You want me to talk to the council?"

"I already told Jess."

"Well you don't need my help then. I need to fetch some food. Seems Ruth's house hasn't been stocked up for a few days."

"Can't you go to the store?"

"What, and face Pam? No thanks, I'd rather give those boys my own rations."

Sarah kissed her mother on the cheek. "Give them my love. Ruth's a good woman."

"She is that." Dawn returned the kiss and hurried towards her cottage.

Sarah stared after her. Her heart was heavy in her chest and she couldn't stop thinking of Martin's face when he'd told her what had happened on the allotment. So grey, as if the life had gone out of him.

She thought of another face, one that hadn't been grey at all. She felt a rush of energy.

She turned and started running.

42

Jess left the house, her heart pounding. Which way?

The coast road was the most direct route. But a storm was threatening, and she had no idea how safe it was. And even then, she didn't know where in Filey she needed to go.

She hurried to the Haywoods' house. She could hear footsteps behind her; was she being followed?

She carried on, not looking round. All that mattered was getting help for Ruth.

At the house, she stopped to hammer on the door.

"I need your help! We have a sick woman. I need to get to Filey."

She could hear voices inside. They stilled at the sound of her voice.

"Please!" she called. She wanted to cry.

She fingered the key in her pocket. The council had been about to give her permission to do this, when Flo and Toni had interrupted.

Not allowing herself the time to change her mind, she plunged the key into the lock and shoved the door open.

Inside was a crowd of men. They turned as one,

staring at her. She counted them: one, two, three... six of them here, just in the hallway. How many more inside?

And where had they come from?

She resisted the urge to retreat, to alert the council.

"I need to find a hospital!"

They stared at her, blinking. They were like a wall of muscle in the confined space. The house smelt of cigarettes and boiled vegetables. She wondered if they had used the rations they'd been given.

"You shouldn't be here," she said. "Where's Mr Haywood?"

"That's me," one of them said. He was the smallest of the men, about five-foot-six with a bald patch and teeth that looked as if they hadn't seen a dentist for some years. "These are my sons, and two of my nephews."

Nephews? No one had said anything about nephews.

"I need your help," she said, swallowing the urge to ask questions, to make accusations. Did they know what had happened to the Bagris? Could the other family be here?

The men parted and a woman pushed though. "What's going on?"

It was the woman she'd met yesterday. She wore a threadbare dressing gown and a pair of slippers with rabbit teeth on the front.

"Oh. You again. How did you get in here?"

"Sorry. I need a hospital. Is there one in Filey?"

"Hah!" the woman spat. She smelled of gin. "You've got to be kidding."

"So where is the nearest one?"

"Fuck knows," the woman replied. "Leeds, for all I know. Wouldn't trust them even if I knew where it was."

"Please," breathed Jess. "My sister is dying. I need to get help for her."

"I thought you had a doctor."

"She *is* the doctor."

"That's unfortunate."

The men sniggered. Jess stared at them, her chest tight. She should leave; this was useless.

"Go to the police," the woman said. "They've got cars. They can get her to the hospital."

"The police?" Jess felt a shudder run through the group of men; she suspected they didn't have a friendly relationship with the local police force.

"Yeah. You'll find them overlooking the beach. Bastards, the lot of them. But they'll help the likes of you."

"Thanks."

The woman jabbed a finger into her chest. "Now get out of here, before we notice the way you broke in."

43

When Jess left the Haywoods' house, there was a growing crowd of villagers outside. They were staring at the house, as if daring its occupants to come out. The graffiti stared back at them.

Clyde was at the back. "What's going on?" she asked.

"They've got wind of what's happening in there."

"Nothing's happening in there." *Not yet.*

"How many of them are in there?" he asked.

"You know that. I saw you outside the council meeting. But how do all this lot know?"

More people arrived at the edge of the crowd. Clyde shrugged.

"I can't deal with this right now," she said. "I have to get help for Ruth. But you're the most responsible person here, so it's your job to disperse this crowd."

"I can't do that."

"You can. We can't have any more trouble. Keep an eye on the house, but don't let anyone attack it."

"What else do you suggest?"

She sighed. "I don't have time, Clyde. Sort it, OK?'

She'd never spoken to him like this before. Clyde and she had an easy relationship, a flirtatious one, at least on his side. Or they had before she'd fallen in love with Zack.

It was dark now, and starting to rain. The quickest way to Filey was along the coastal path. But in the dark, that would be foolish. The roads would be quiet, and possibly lit once she got closer to the town, although she doubted it.

She gave Clyde a look that she hoped meant business then headed for her old house. There was a bike round at the back, one she'd occasionally used to get around the village. Sarah had taken it to Filey when Martin was under arrest. And Ben had cycled to the farm in search of Ruth.

She ran to her own house, hoping the new occupants wouldn't see her. Two days ago it had still been propped up behind Jess's bin.

She ducked and slid around the side of the house. There was a window here, to the side of the living room, but it was high in the wall. There was no sound from within, and no light. Lights-out had been and gone. She thought of Ben and Ruth, in their house with no light and no heat. She should have lit a fire.

She should go back. She didn't know the specifics of how to override the power settings, but she had the authority to make it happen. At this time of night, the generator would be switched off. Only an hour's power per day, and that just after dusk.

But with Ruth almost drowned, no one would say no to Ben if he asked the same thing. She just had to hope he'd have the presence of mind. Or that Colin would.

Focus, she told herself. *Get help*. She dragged the bike out from its hiding place, wishing that she'd ridden it at some point while she'd been living here. It was pitted with scraps of rust from being left out in the rain, and there was mud etched onto the frame. But it would be usable.

She pulled it onto the road, keeping as quiet as possible, then mounted it. She hadn't ridden a bike in years and unlike her brother, she'd never been a champion racer. He'd told her how good it had felt speeding to the farm, feeling the wind on his face and catching a moment when he forgot why he was riding and just enjoyed the freedom of it.

She picked up pace and sped to the edge of the village, past the house she'd lived in with Sonia. No one was about; the village was quiet. Those villagers who weren't outside the newcomers' houses were locked away indoors.

At the road, she hesitated. She had no idea how frequently this road was used now. When she'd lived out on this edge of the village she'd listened to the occasional vehicle passing at night, and wondered where they got the fuel from.

She had to risk it. If she heard something coming, she could dive into a hedge.

She started cycling. The road led up a slight gradient and she felt the muscles in her thighs tense with the effort. After a while, the road levelled out. She swept past a gloomy hulk on the left, what might have been an abandoned petrol station or small supermarket, and took a right turn when the road forked. She just had to keep turning towards the sea, surely, and she would find it.

Filey announced itself with a smattering of houses, some with dim light inside and others clearly ruined. She could smell the residue of the explosion: an acrid smell that coated the insides of her nostrils, and a heaviness to the air that made her gag.

She pulled her shirt up, trying to cover her face, and pedalled on. The road went uphill and then steeply downhill again. At the bottom of the hill was the sea. She stopped, wary of the steepness of the hill and the fact that she couldn't see where the road ended and the water began.

She looked around her. The police station overlooked the beach, she knew that much.

The only sound was of the wind ahead of her, blowing in from the sea, and the cry of a single seagull somewhere to her right. Next to her, a large building was set back from the road. She approached it. A pole in front of it had long since lost the sign that would have sat on top.

Then she saw the police sign by the door. She allowed herself a shiver of relief and hurried to it.

44

SARAH KNOCKED ON THE DOOR, trying not to hammer too hard. She shifted from foot to foot; it was cold.

The door opened. A gust of warm air blew out.

"You."

"Let me in," Sarah said.

"What do you want?"

"I want to talk to you. I want to know why you lied."

Sally shrugged. She pushed Sarah out of the way and looked past her. "Where's your boyfriend?"

"Martin's at home. I'm on my own."

Sally crossed her arms and cocked her head. "You should have left him. After what he did."

"They made him."

"*They made him.* You are so stupid."

"Please, let me in. It's freezing out here. And we're letting the rain in."

Sally pursed her lips and looked Sarah up and down. "Alright then." She pulled the door wider and let Sarah pass.

The living room of Sally and Mark's house was lit by

just two candles. They flickered into the corners, sending long shadows up the walls. Sarah wondered why they didn't have any more.

"Where's Mark?"

"In bed."

Sarah glanced up the stairs. No sound came from above. There was no reason to assume Sally was lying; where else would he be?

She walked further into the room and turned to face the other woman. Sally was right behind her, her breath hot on Sarah's cheeks as she turned. Alarmed, Sarah took a step back and almost tripped over a low table.

"Better be quick," Sally said. She sat down on a threadbare sofa. She didn't invite Sarah to do the same. "You've come to apologise, I hope."

"Why would I do that?"

"Because of what Martin did to my Mark. Have you seen his face?"

"Yes. But there's no evidence it was Martin."

Sally rolled her eyes. "He kidnapped us. He got himself arrested. Who else would it have been?"

Sarah shrugged. "It wasn't Martin."

"What's going on?"

Both women turned to see a man coming down the stairs. He was short and slim, with dark, curly hair and a scruffy beard. Sarah tried to remember if she'd ever seen him before.

"Mark darling, I thought you were asleep." Sally's voice had changed; it was no longer sharp, but honeyed.

He rubbed his eyes. "I heard voices. Everything OK?"

"Sarah's just leaving."

Mark looked at her. "Good."

Sarah stepped towards him. "Can I look at your face?"

He put his hand up to his cheek and cast Sally an anxious look. "I'd rather you didn't."

"Please. I want to see what happened to you."

Again he looked at Sally. "Your boyfriend attacked me, is what happened." His voice was low.

Sarah leaned towards him. He looked at Sally again and she nodded. Slowly, he dropped his hand.

He had a thin line running diagonally across his cheek, and another over his eyebrow. The skin around his eye was purpling, with yellow tinges at the edges.

"You say Martin did this?" she asked.

"He did." Mark didn't meet her eye.

"Why?"

"What do you mean, why?"

"Why would he attack you? Why you, in particular?"

He shrugged. "Who's to know I'm the only one?"

"Leave him alone." Sally rose. She placed herself between Sarah and Mark. Her face was in darkness. "Stop harassing us."

"I'm not harassing anyone. I just want to know the truth. Why would Martin attack your boyfriend?"

"Husband."

"I don't remember any wedding."

"That's none of your business."

"Look, just tell me why you think he did it, and I'll leave you alone."

Sally and Mark exchanged glances. They said nothing.

"It couldn't be because you threw a brick through our window, could it? Or because you posted shit through our door?"

Sarah listened to herself. What would Dawn say, if she heard her saying *shit*?

Mark flinched. His eyes darted to Sally's face. She was staring back at him, her back to Sarah now.

"I'm right, aren't I? It was you who did those things. Why do you want him dead?"

"He's trouble," said Sally.

"You don't know what you're talking about," said Mark. "I didn't do anything of the sort."

"No, but maybe your *wife* did."

Sally turned. "Get out," she spat into Sarah's face. Sarah shrank back.

"Happily. I don't want to stay in this sick household any longer than I have to."

Sally came after her into the night. "If you think we did that, then you know why he attacked Mark. You can't deny it!"

Sarah ignored her and walked away. Sally had a point. But there was something about those injuries that didn't add up.

45

The door to the police station was unlocked. Jess pushed it open and felt the warm air hit her. She hadn't realised how cold she'd been.

Inside was a bare waiting room, two threadbare chairs against one wall and a long desk on the other side. There was no one sitting at it.

She walked to the desk and looked for a bell. There was nothing. A paperweight sat in the middle of the desk, two manila files one on top of the other and a pencil case that bulged with pens. She thought of the way writing materials were rationed in the village. Paper and pens couldn't be grown, they had to be bought with what limited money they had, and took low priority beneath medicines and food that couldn't be grown.

She put her hand back in her pocket; she was no thief.

A man emerged from the door at the back of the space. When he saw Jess, he stopped.

"Can I help you?"

"I'm Jess Dyer. From the refugee village down the coast. We need help."

He frowned. "You're not the only one."

"Sorry?"

"We've got another one of you here."

She frowned. "Who?"

"Can't tell you. Sorry."

"I'm the steward. The leader of the village council. If one of our people needs help, then I should know."

"You don't have any jurisdiction here. You do know that your little council means nothing, don't you?"

She sniffed. "Not to us. It keeps our community stable."

"I'm sure it does. I remember you. Your sister-in-law was arrested. And you assaulted Anita Chopra."

She shifted her feet. "My sister-in-law is the person I've come about. She needs an ambulance."

He raised an eyebrow. "This is a police station."

"Is there a hospital in Filey?"

"Of course not."

"But can you request an ambulance for Ruth?"

"I thought you looked after your own? You didn't even want us to investigate the alleged murder of your fiancé."

His tone made her feel like she had when she'd hit Anita. This was a bad idea; the police would never help them. But she had to try.

"Ruth is the village doctor. She almost drowned today."

The eyebrow rose further. "Really?"

"Don't you believe me?"

"Look. Wait here. I've got things to deal with. I'll be back shortly."

"She can't wait. What if she—?"

But he was gone.

Jess threw herself into one of the chairs. The space felt musty and institutional, taking her back to dentists' surgeries and hospital waiting rooms before the floods. A solitary lightbulb hung from the ceiling, casting a dull yellow glow over the space.

Should she wait, or was there somewhere else she could go?

It was late. Even if Anita Chopra had an office in Filey, it would be long since closed. And there was no hospital. The police were her only hope.

She went to the glazed door that the man had disappeared into and peered through. Beyond it was a dim corridor. Someone moved at its opposite end, too indistinct for her to make out.

She grabbed the door handle and tugged. It was locked. She went back to the reception desk, wondering if those files would help her.

She picked one of them up, then heard movement behind her. A young woman emerged from the other door. She eyed Jess, then went through the same door as the man. Jess held the files behind her back. When the woman was out of site, she dropped them back onto the desk.

Her heart was pounding. Every time she thought of Ruth, she felt a vice grip her chest. She had to help her, and fast.

46

Sarah ran home, her senses alert and her feet light. She felt nothing more than anger: bright, energetic anger, lifting her up and throwing her along the roads.

Sally and Mark were lying, she knew it. There was something about the tone of Mark's voice, the way he looked at his girlfriend. And he'd been careful not to actually say that Martin had put those marks on his face. Sure, Sally had repeated her accusations, but Mark himself had not.

So if Martin hadn't attacked Mark, who had? Why? And why would he lie about it?

Maybe Martin could provide some answers. Someone had to.

She reached Martin's flat. She divided her life between his home and her mother's, never really thinking of either as her own. At some point, she would have to choose. Her mother was independent now, a member of the village council. But Sarah's urge to protect her, at least to be with her if she couldn't shield her, was strong.

She unlocked the outer door, slid inside so as not to

disturb the neighbours, and locked it behind her. She took the stairs as quietly as she could and unlocked the door to the flat.

"Martin?"

No response. Was he in bed? She felt irritation wash over her.

"Martin?" She pushed open the door to the bedroom.

The bed was empty. The duvet was piled in its centre. Martin's jeans, which had been folded over a chair, were gone.

She turned back to the living room. "Martin? I'm back." But he wasn't there.

She opened the door to the bathroom; also empty. She stepped back, suddenly scared.

She went to the sink and poured a glass of water. She gulped it down and wiped the spillage from her shirt.

She went to the window and looked out. The village was dark now, and silent. Not even the sound of wildlife. It was as if she'd been plunged into a thick soup of blackness.

She tore off the piece of wood that Martin had used to cover the broken window. A blast of cold air blew into her face, making her blink. She resisted the urge to call his name out into the night.

Martin, where are you? He knew as well as anyone that there was a curfew. And he also knew that someone in this village wanted to kill him.

She went to the door, hesitating before opening it. He could be anywhere. Maybe they'd been attacked again and he'd gone to help. But she hadn't heard anything.

She looked around the room. His coat was gone, and his shoes. There was a mug next to the sink which had been washed out. She was sure it hadn't been there before.

So he'd meant to leave. He'd had the time to grab his

coat, and wash his mug. He hadn't been snatched. That was something.

There was a piece of paper in the seat of the armchair next to the window. She picked it up. It was the same paper that had been thrown through the window: *Leave, or die.* Any normal person would throw a note like this away. But maybe Martin thought it held a clue.

She turned it over. There was writing on it, in faint pencil. She pulled it up to her face, but it was too dark to read.

She went into the kitchen and lit a candle. She held it near the note, careful not to make contact.

Wait here. M xxx

She felt her limbs weaken. Where had he gone? Was he going to confront his attacker? And should she stay here, like he asked, or go looking for him?

47

"RUTH. Ruth, it's me, Ben. You're at home. You're safe."

Ruth felt as if her chest would explode. She reached inside herself to identify the source of the pressure, but could feel nothing. She gulped in air and coughed, feeling like she might choke on her own breath.

She felt a hand on hers, tight and hot. She tried to move her arm away but it was too heavy.

She moaned.

She felt lips brush her cheek. She blinked, frozen. Where was she? Who was this person, touching her? *Leave me alone*, she thought. *Let me go.*

She clenched her eyes shut. Maybe if they thought she was unconscious, or even asleep, they would go.

She focused on her breathing, keeping it calm and steady. *Don't draw attention to yourself.*

The hand loosened on hers. She felt herself lighten.

The pain in her chest was hard and rough. It extended from deep inside and rasped its way up her windpipe to her lips, feeling like it might tear through her flesh. Her legs felt heavy and dull, her skin hot.

She worked her mind over the parts of her body, trying to regain sensation, to take control of herself. Feet first; aching and sore. Then her legs, heavy and loose. Her torso; she skipped that, overwhelmed by pain. Her arms felt hot and light, totally unlike her legs. She felt as if they might fly up into the air if she let go of them.

She moved up to her neck. It was stiff and cold. Her head felt as if it was packed full of burning cotton wool. Pain jabbed at her eyes and ears, threatening to pierce right into her, to tear her head apart.

She moaned again.

"Ruth?"

This person was never going to leave. Not until she told them to.

She fluttered her eyes open. Above her was a dim space, light flickering off to one side. A shape hovered over her. She blinked and the shape came into focus; it was a person.

She gasped, trying to pull herself down and into the bed. Robert! She had to get away.

She felt her legs convulse, as if trying to run from the bed. That was it. She had to run.

She let her legs fall to one side, feeling for the edge of the bed. At last she felt the mattress fall away. She shifted her weight, knowing that if she wasn't careful, she would fall. And then she would be powerless.

"Ruth? What are you doing?"

She muffled a scream. He put a hand on her chest and she tried to pull her muscles in, to create a cave of her flesh so he couldn't touch her.

She pushed air up onto her mouth in an attempt to speak. Nothing but a hoarse croak came out.

She opened her eyes again. The figure had pulled back and the hand was off her chest.

"She's scared, Ben. Maybe we should leave her alone for a bit."

"I can't leave her alone. I have to watch her."

She felt that icy fist clutch at her stomach again. Robert Cope was in here, watching her. He wasn't going to let her go. Not ever.

She pushed out a shaky breath. "Get off me," she whispered.

He leaned over her. "Ruth love, it's good to hear your voice. What happened?"

"You took me," she whispered. "You know what happened."

"I didn't take you anywhere, love. I brought you home."

The back of her throat rasped. She coughed.

A hand looped around her back and pulled her up. She felt herself being raised in the bed, horrified. Her legs kicked out and she screamed.

The hand pulled back and she dropped to the bed again. *Thank God.*

"It's me, Ben. I've got you."

"We want to get you better." Another voice, female. Who were the other women they were holding? She couldn't remember.

"Leave me," she croaked.

"Come on, Ben. Let's leave her for a bit. She needs time. Maybe the cold did something to her senses."

"I'm scared, Sheila."

"I know, sweetheart. But we're scaring her. You go downstairs to those boys, and I'll keep an eye on her from the door."

Ruth felt her breath become more steady. If one of the other women was watching over her, then maybe she'd be safe.

48

Jess paced the room. From time to time, she stopped at one of the doors and tried to open it. She pushed her face against the glass and peered into the dimly lit space, leaving a mark where her breath hit it.

She wondered if there was supposed to be a receptionist of some sort at the front desk. She examined the items on the desk; the paperweight was dusty but the pencil case and file looked recently used. It seemed like the desk had been abandoned, or maybe that it had been left temporarily by someone who had decided – or been told – not to return.

Would Ruth have woken up yet? She tried to imagine how Ben would be when his wife came round. She hoped he could keep control of his emotions.

"Ms Dyer?"

She span round to see a woman standing in the open doorway behind her. It was the young woman who'd passed her before.

She hurried to her.

"I'm Detective Constable Paretska," she said.

"I remember you. You were there when…"

The woman nodded.

"Has he told you what's happened?"

"Yes. We called the hospital. But there's a complication."

Jess felt her heart sink to her shoes. "What kind of complication? Is Ruth alright?"

DC Paretska blinked and looked away. "I'm sorry, but I don't know. You don't have phones in your village."

"Of course we don't." *Because we're only allowed a certain amount of power. Because we don't have the money. Because people like you think it's inappropriate for refugees to have luxuries like phones, and confiscated them when we arrived.*

"Have they sent an ambulance?"

"Come with me."

Jess pushed down her anger and impatience and followed the woman through the door and along the corridor she'd been spying on. She began to wonder if she'd made the right decision coming here; this was getting them nowhere.

DC Paretska stopped at a closed door. She opened it and ushered her through.

Inside were three people. The man she'd spoken to before, who she remembered as the Detective Sergeant who'd arrested Ruth before. Anita Chopra, wearing a crumpled jacket and with mascara rimming her eyes. And Martin.

49

RUTH SAT up in the bed, wincing as Sheila plumped cushions behind her.

"Here you go," Sheila said in her sing-song voice. "Let's make you comfy."

"Where am I?" asked Ruth.

Sheila stopped moving. "Why, you're in your own bed of course. Don't you recognise it?"

Ruth looked around. The room was a blur, dancing shapes and patterns of light and dark. She squinted to try to make out details, but all she could see was a white mist.

"I don't feel so well."

Sheila wheezed as she lowered Ruth to the pillows. Ruth felt the softness of them hit the back of her head. She tensed, feeling as if she might be sucked into the bed.

Sheila's hand was on her cheek. "You look peaky, love."

Ruth tried to move her head but it wouldn't cooperate.

"Ruth? Sheila, is it alright to come in?"

Ruth squeezed her eyes shut. It was him again.

The room filled with hazy figures, people moving around. She tightened her eyes and tried to shut it out. She could just about cope with Sheila here; the older woman had a confidence that made Ruth feel more solid than she had, as if she could cling onto the edge of the world and not fall off. But now there were more people: moving, swaying, talking.

Sheila bent to her. "It's just Ben, love. Your husband, remember? And your kiddies. Little Sean and Ollie." Her voice dipped to a whisper. "They're scared."

Ruth felt fear grip at her. What were her boys doing in a house with Robert Cope in it? She needed to get them to safety, quickly.

"Take them away, get them away. They aren't safe!" she whimpered.

"It's fine, Ruth. They're with me." A man, on the other side of the bed. Sheila had said it was Ben. He smelled familiar, of soap and cooking fat. She tried to remember Robert's smell: aftershave, wasn't it?

She sniffed. She couldn't smell any aftershave. Maybe this wasn't him. Maybe her boys would be safe, after all. She opened her eyes.

A man bent over the bed, his eyes boring into her. She raised her hand to shield herself from his gaze, and shifted her head to the side.

The sight of two small boys snagged on her vision. She gasped. Blonde hair; one neat, the other tousled. She knew them.

She felt a tear roll down her cheek. "Boys."

The boy on the right – Sean, she was sure of it – nodded. His brother turned away. She felt her stomach dip.

"Ollie. Ollie, come to Mummy."

The boy turned back to her. His lip was trembling, and his eyes were red. He shouldn't be here.

"Take them away," she said. She wanted to pull them

to her, to clutch them so tightly that none of them would be able to breathe. But she didn't trust herself with them. She didn't trust that in this room, Robert wouldn't return. That he wouldn't come back for them.

She fell back against the pillow, raked with sobs. Sheila muttered something and the man she hoped was Ben bent over her again. He laid a kiss on her forehead. She stiffened, holding herself still. He grabbed the two boys by the hand and led them out.

Sheila stroked her arm. "Well done, sweetheart. How did it feel, to see your boys again?"

Numbness was creeping up her legs. There was a deep pain inside her abdomen, growing and threatening to spill out of her. She scratched at her skin, frightened.

She stared up at Sheila, willing her to come for her, to come inside the hole Ruth had fallen down. *Save me*, she thought.

The pain jabbed at her, then spread up her chest, bringing numbness behind it. She felt herself sway against the pillows and then the world went black.

50

"MARTIN?"

Martin stood up and gave Jess a nervous smile. "Hi, Jess."

"What are you doing here? Where's Sarah?"

She looked from him to the two detectives and the council officer. Had they brought him here?

"Have you been arrested?" she asked him.

He shook his head. "No. I came here myself."

"Why?"

Martin bowed his head. "I wanted to help."

Anita Chopra stepped forward. "He's offered to do an exchange."

"What sort of exchange?"

"I gather he's been allocated a flat, in your village. He says we can house people in it. He'll move to the hostel we've established, here in Filey."

Jess stared at Martin. "That flat isn't yours to give them. I worked bloody hard to get you—"

"Nobody wants me there," he said. "I've been getting death threats."

"I know. It doesn't mean anything."

"I can't put Sarah at risk."

"Does she know you've done this?"

A blush. "No."

"And is she going to live here with you, or go back to Dawn?"

"She officially lives with Dawn still. But I hope we can work something out."

"She won't move here."

He shrugged. "I had to protect her."

She gritted her teeth. She'd fought to protect this young man's rights, to persuade the other council members and then villagers to allow him to live among them. She'd thought his acceptance showed something about the village, and what it stood for. But now, at the slightest perceived threat, he was running away. Maybe she was wrong about him.

"I haven't got time for this," she said. She turned to Anita Chopra. "Ruth needs help. She almost drowned today. I left her at home, I have no idea if she's going to be alright."

A shadow crossed Anita Chopra's face. "You haven't come here to assault me again?"

"I'm sorry about that. I was grieving. I was angry. I took it out on you."

"You did."

"So?" asked Jess. "Can you help her? I've already asked the police if they'll call an ambulance." She turned to the detectives. "Have you?"

Anita stepped towards her. Jess gave her a *back off* look, but she ignored it. She flicked her gaze to the woman's chin; there was no scar.

"I thought you looked after your own," Anita said.

"The only doctor we have can't treat herself. We need help."

She stared back at the official. She hated having to come to them like this, to admit weakness. But Ruth's life was at stake.

"We can help you, yes."

Jess let out the breath she had been holding. "Thank God. She's at home. Tell them to hurry."

"But we need something from you in return."

"What kind of people are you? There's a woman who could be dying, and you won't send her an ambulance?"

"You don't pay taxes. You don't contribute to the wider community."

"We pay our way. We send men out to work on the earthworks."

"Casual labour. Hardly a full role in society, would you say?"

"I don't see what that has to do with Ruth!"

Anita sat down. She gestured for Jess to do the same. Jess ignored her.

"Jess – can I call you Jess?"

Jess nodded, holding in her anger and frustration. She wanted to claw this woman's eyes out.

"Jess, you know how society works. You make what contribution you can, and the state helps you out when you need it. Medical help, housing, that sort of thing."

Jess narrowed her eyes. "You left us to survive alone."

"No. You chose to. Don't you remember what that village was in the early days?"

"I don't know what you're talking about."

"Hmm." Anita raised her eyes to the ceiling, as if calculating something. "No. You arrived just as we were leaving. You missed the worst of it."

"I still don't know what you're talking about. Or why it means you can't help Ruth. The village needs her."

"The place was lawless. People who'd spent months living rough, eking out a living on the road. People get used to different standards in those conditions, I imagine. We let you stay, on the understanding you would govern yourselves, and that you wouldn't cause us any trouble. That you wouldn't be a drain on resources."

"That was six-and-a-half years ago."

"Just under six, to be precise. But yes. And now you expect us to help you, but you won't help us."

"Is this why you sent more people, without telling us?"

Anita frowned. "I'm sorry?"

"There are almost twenty people in those houses, that shouldn't be there. They say they're just looking for shelter, but"—she glanced at Martin—"how can we be sure?"

Anita licked her lips. "If there are people squatting, then we'll deal with it."

Jess thought of the little girl she'd seen, sitting on the stairs of the house. How old had she been – four, five? She remembered the children she'd seen sleeping on the floors of church halls and sheltering in hedgerows after the floods.

The anger left her.

"You will?"

"Of course. We want to make sure everyone is looked after, not just those who are prepared to break the law."

Jess looked at Martin again. He was blinking at her, nodding. Nothing was what it seemed.

"Tell me what you need," she said.

Anita stood up. "At last. We understand each other." She scratched her chin. "It's simple, Jess. You let us make better use of the hosting stock in that village, and we help your sister-in-law. Say the word, and I'll make a call."

Jess thought of Ben, adamant that they weren't taken over by outsiders. Of Ruth, and her fear of everyone who wasn't familiar. Jess had failed to notice Ruth's deterioration before her eyes.

Then she thought of Zack, lying in her arms, bleeding to death. Of the people who'd invaded their village.

"They killed him," she whispered.

"I'm sorry?"

"My fiancé. Zachary Golder. He was murdered by people from your so-called civilisation."

"When was this?"

"Two nights ago."

"And what did you do about it?" Anita looked towards the police officers. They shook their heads.

"Of course we didn't tell the police," Jess said. "They never would have listened."

"Don't you see why?" Anita said. "You can break this deadlock. You can get medical help for your sister-in-law. Justice for your boyfriend, maybe. Just say the word."

She thought of the village meeting, the way she'd been trampled on, ignored. The helplessness she felt.

This wasn't her decision to make. The council was a democratic body, with clear processes. The village had survived all this time by allowing everybody to have their say.

But Ruth could die. Ben would see that this had to be done. Wouldn't he?

"Alright," she muttered.

"Sorry?"

She jerked her head up. "I said yes. Send her an ambulance. Ask these two to help find Zack's killer. We'll work with you on the housing."

51

SARAH RAN out into the night. She turned wildly from side to side, knowing that expecting to find him nearby was just wishful thinking.

How long had she been gone? It had turned dark since she'd arrived at Sally and Mark's house almost an hour ago. And it had been dusk when she'd left Martin. He'd been lying in bed feeling sorry for himself. It hadn't occurred to her for a moment that he'd go, much less that he'd do something stupid.

She ran back to Sally and Mark's house, her agitation rising the closer she came to it. If he'd gone to confront them, who knows what recklessness might have come over him?

She stopped in her tracks, almost falling over in her shock.

What if Sally had been right? What if Martin *had* discovered that it was Mark who'd targeted him? What if he'd decided to retaliate? Martin could be violent; she'd seen him hurl himself at Robert Cope. He'd claimed to be protecting her, but what if it had really been revenge that motivated him?

If he could plunge a knife into a man who'd once saved his life, what was he capable of doing to a man who'd posted human shit through his letterbox?

She was slumped on the ground, grass scratching at her bare legs. She hadn't even pulled the door to the flats closed behind her. Their neighbours wouldn't forgive her.

She had to know. She had to face him.

She squared her shoulders and pointed herself in the direction of the house. The buildings around her all looked the same; dark, squat lumps of stone or brick, looming around her. Which way had she come from? And which way did she need to go?

She closed her eyes for a moment, waiting for the dizziness to recede. When she opened them, she was back in the familiar village. She knew where she had to go. She started running.

At Sally and Mark's house, she hammered on the door, not caring this time about being subtle. The door flew open almost instantly.

"I've told you to fuck off!" Sally's eyes were blazing.

"Is Martin here?"

"No. Why the hell would he be?"

"I just thought he might have come here."

Sally pushed the door against her. Sarah caught it with her foot. "Please. He's disappeared."

"Good."

Sarah dug her fingernail into the flesh of her palm, anxious to conceal her panic and frustration. "If you see him, will you let me know?"

"What's going on?" Mark was behind Sally, his face shadowed in the dim light of the candles.

"It's this bitch again."

"I'm looking for Martin. I thought he might have come here."

Mark and Sally exchanged glances. "No," said Mark. "He's buggered off, hasn't he? Like he should have in the first place."

Sally reached round to touch Mark's face. "It was worth it, then."

Mark gave her a look and retreated inside. Sally took advantage of Sarah's surprise and slammed the door on her, narrowly avoiding her toes.

Sarah stood and stared at the house, confused. What was going on with that pair? And where was Martin?

She looked back towards the village centre. Her chest was heaving now, her breaths shallow and fast. She didn't know if she was thinking straight, or if panic had taken over. Martin had left the village before, when he thought she was at risk from his presence. She'd gone with him that time, defying her father. But this was different. He'd lived here for six months. He had a job on the allotments. He came home at night talking to her about the chicken pens they'd built, about the personalities of the different chickens. It was almost as if he saw them as pets; she'd worried he might baulk at eating them when the time came.

He had everything to stay here for, and nowhere to go. He'd never go back to that farm, and Filey just held memories of his arrest for Robert's murder.

She pulled at the skin on her face, berating herself. Maybe he was hiding out on the allotments like he had when her father had been searching for him after he'd first arrived.

She turned towards them.

She picked her way between the rows of vegetables, anxious not to trample the precious food. There were two sheds in these allotments: one on this side and another at the far edge. She hurried to the closest one, and tore the door open.

It was empty. A spade leaned against the wall, a pile of flowerpots teetered in a corner, and cobwebs strewed the ceiling. She shuddered and closed the door again.

The other shed was equally deserted.

Maybe if she tried her mother's house. Dawn and Martin had grown closer over the last six months. He hadn't wanted Dawn to know about the threats, but maybe he'd changed his mind.

She considered going back to the flat, just in case. But it was in the opposite direction from the clifftops. And she needed Dawn's reassurance that she wasn't going mad.

She made for the cottage.

She started to run.

52

JESS SAT in the back of the car, feeling uncomfortable. The rough fabric of the seat scratched through her thin jeans and the stale air jabbed at her nostrils. She realised this was the smell of civilisation, and she didn't miss it.

They swept along the main road down the coast, headlights picking out bushes and crumpled road signs as they careered round the bends. She gripped the seat front with her ragged fingernails, fighting nausea.

Martin was beside her in the back. In the front was Anita Chopra, and a uniformed police officer. He stared ahead in silence, all his senses focused on the road ahead.

At last they reached the entrance to the village. The car slowed and took the corner steadily. She turned to see the car behind follow them. The two detectives were in there, along with another uniformed policewoman. And there was a dark van behind that; she'd asked Anita what it was for, but had got no response.

She stared out at the houses as they glided along the Parade. People inside would be coming to their windows, wondering why two cars were arriving at the village at this time of night. Both cars were unmarked: she'd managed to persuade the senior detective that the villagers didn't need the shock of police cars descending on them two days after a vicious attack, and that Ruth in particular didn't need to be reminded of the last time they'd been here.

They reached the circle of grass that marked this edge of the village centre. She stared at the oak tree on its opposite side, her vision clouded. She'd seen Zack run out in front of that tree, heard him cry out her name. Then, nothing. Could she have saved him? Even if she hadn't been able to stop the attack itself, could she have been in the right place, at the right time?

She closed her eyes. She had to focus on Ruth. She didn't need to be grieving for two people she loved.

"You know the way?" she asked, superfluously. Of course they did. Every time the authorities came to this village, they made straight for Ben and Ruth's house.

The car stopped outside the house. The front door opened and Ben emerged. Jess climbed out of the back of the car and steeled herself. The van had gone; where was it?

"What did you do?" he cried.

"Is the ambulance here?"

"What ambulance?'

"They called an ambulance. Has it arrived yet?"

'I don't know what you're talking about, sis. Why did you bring the police here?"

She grabbed his wrists. He pulled against her grip but she held tight.

"Listen to me, Ben. I persuaded them to send help. An ambulance will be here any minute. How is she?"

"You've been gone for hours."

Not hours, she thought. *But too long*. She felt panic rise in her throat. "How is she?"

He stared into her face. "Not good."

"Is she conscious?"

"She has been. But she's asleep. I think. I hope."

Jess looked towards the house. "Can I see her?"

They both turned at the sound of advancing sirens. Blue lights reflected off the windows and glowed through the gaps between buildings. This would escape no one's attention, thought Jess. They'd want to know how she'd persuaded them to help.

The ambulance pulled up next to the car. Two paramedics jumped out. One of them, a grey-haired woman, spotted DS Bryce and hurried to him. The detective pointed at Ben and Ruth's open door and the two paramedics hurried inside.

Jess wondered what it would have been like if they'd had access to professional medical care when Sonia had been sick. When Zack had been stabbed. Why hadn't she gone to the authorities years ago?

Ben ran after the paramedics, calling instructions.

"Leave them to do their job," Jess called. He ignored her.

Sheila came out of the house, her eyes wild.

Jess stepped forward. "Thanks for looking after her. How is she?"

"Not good. Ben thinks she's asleep but it didn't look like that to me."

Jess felt the blood leave her face. "She's not—?"

Sheila put a hand on her arm. "No, love. She's still with us. But she's going in and out of consciousness, like she's struggling to hold on. At one point she started shouting a name."

"A name?"

Sheila leaned in towards her. "Robert."

Jess stiffened. "Are you sure?"

Sheila nodded. Jess looked back at the house. Why was Ruth shouting Robert's name? Did she think he was there with her, in the house? Did she think Ben was him?

She shivered. "Where are the boys?"

Sheila sighed. "Asleep, on the sofa. Little loves."

"I'll look after them now."

"Surely you need to deal with all these people."

"They can wait."

Sheila smiled. "You're right, love. Family first."

Jess strode towards the house. She was halted in her tracks by the paramedics, pushing Ruth out on a trolley. She stood aside to let them pass, then watched as they loaded her onto the ambulance. Ben trailed behind them, climbing up into the ambulance to be with his wife.

"Where will they take her?" she called out, not sure if anyone who might know was listening.

"I'll ask them, love," said Sheila. "You go in to those kiddies."

Jess nodded and stepped into the house.

53

Ruth stared out of the window. All she could see was a pale sky, and the tip of a mast on the building opposite. She wondered how far they were from the sea.

A nurse bustled in, carrying a clipboard. "Hello, love. How are you feeling?"

Ruth said nothing, but gave her the best smile she could muster.

"Your husband's here."

Ruth felt her skin constrict. This hospital was so anonymous, so removed from everything that had happened to her. None of these people knew her or what she had done, what had been done to her. She could dissolve in the anonymity of it, the routine. She could let herself be carried on it until maybe, just maybe, she came out the other side.

"I don't want to see him."

The nurse stopped in her tracks. "You sure?"

Ruth nodded. Her chest felt heavy and there was a stillness behind her eyelids that felt like her head had been filled with glass.

"I can't."

The nurse perked up. "We won't force anything on you. I'll tell him to wait."

Ruth slumped against the pillows. "No. It's alright."

If she didn't see Ben, then she wouldn't be able to see Sean and Ollie. They needed to know she was getting better.

The nurse smiled and went away while Ruth took deep breaths and ran her fingers through her hair repeatedly until Ben appeared. He was smiling sheepishly.

"Hey, love."

"Where are the boys?"

"They're at home, with Jess. I thought this would be too much for them."

She closed her eyes. She shouldn't have let him come. She couldn't stand it.

She opened her eyes again. She forced herself to explore his face, to take in every aspect of him. *He's not Robert*, she reminded herself. *He won't hurt you.*

But it was all his fault.

"I want to see the boys."

"I can bring them tomorrow."

She nodded, her body feeling as if it was full of nail filings. She forced out a smile.

"How are you?" he asked.

"I'm fine," she lied.

"Good."

She widened her eyes. Did he believe her? Was he that insensitive?

"How did I get here?"

He shuffled his orange plastic chair towards her. She gave him a warning look and he stopped moving.

"Jess went to the police, in Filey. She found out where

the station was from those people they moved into the village."

She swallowed. Her throat was sore and her stomach felt like it had been turned inside out. "Are they still there?"

"Who?"

"The new people."

"No. Turns out they were hiding some family members, who were wanted by the police. They arrested some of them, and threw the others out."

She felt fatigue wash over her.

"Good riddance," Ben said. "We don't need outsiders."

"They won't all be like that."

Ben sat upright. "What? You didn't want them here any more than I did."

She swallowed down her distaste. *You never listen.*

"The explosion. They were all made homeless. Just like us, with the floods."

Her voice was hoarse; she needed to protect it. She grabbed his hand.

"Come back tomorrow, with the boys. And tell Jess she needs to do what that council officer says. It'll be worse, if she doesn't."

54

"You had no authority."

Ben glared at Jess. His sons were still asleep. Anita Chopra sat at the third kitchen table, perfume surrounding her like a cloud.

Ben had spent the night at the hospital with Ruth. This morning, Anita had brought him back to the village. Ruth was awake and responsive. She was calling him Ben instead of Robert. But she didn't want him there. So here he was, losing his rag with Jess again.

She leaned over the table, staring him down. "I had no choice."

He looked from her to Anita. True to the promise she'd given Jess last night, she'd said nothing of their deal on the way here with Ben. That had been left for Jess to do.

"We'll overturn it," he said.

"You can't."

"We can. We will. Ruth is in that hospital because of what happened the last time we welcomed outsiders into this village."

"She's in that hospital because we were too busy

fighting to see how sick she was. *And* because of what you and Robert Cope did when you were teenagers."

He reddened. "How dare you."

"You know what you did." She glanced towards Anita. "You said it yourself. And it brought Robert Cope here. Look at what you've done to Ruth."

"I didn't do that."

She felt as if he'd punched her. "Is that the best you can come up with?"

"I looked after her, Jess. While you buggered off looking for help."

"From what I hear, it was Sheila we have to thank."

He reddened. "What do you expect me to say?"

She tried to drag some energy out of her body. She was tired. She missed Zack.

"We need to accept help. Whatever the terms. The people from Filey, they're just like we were once. People who've lost their homes because of a disaster. Don't you remember what it was like to feel like that?"

"They attacked us. They've victimised us for the last six years."

"I know. But why does that mean we have to be as bad as them? Can't we be the ones to forgive?"

"Have you met them? Do they want to make friends?"

She thought of the men in that hallway. The way they'd crowded around her.

But they were gone now. The Bagris were also gone, too scared of the neighbours who'd been forced on them. She wondered if they had family to stay with in Filey.

But new families would soon arrive in their place. Families who weren't intent on trouble. More than two of them.

"What does Ruth think?" she asked.

"About what?"

"About us letting new people come here. About this village opening itself up to the outside world."

He reddened. "You know what she thinks."

"Tell me."

"She's Ruth. She wants to help them, of course."

"Don't you think she's right?"

"She's not a council member. It isn't up to her."

She felt her mouth drop open. "She's your wife, Ben. And she's the most sensible person I know."

His shoulders dropped. "I feel like I've lost her, sis."

"Well, you have to win her back then. She's been through a lot. She's been very sick. It's not going to be easy, and you have to be patient."

"I know."

"And not just with Ruth."

"Huh?"

"The new people. If we make them feel welcome, they'll behave themselves."

"How do you know that?"

"Because I know what this place was like, before we arrived. Anita told me." She exchanged glances with the council officer. "It calmed down. It can again."

He shook his head, staring at her like she was an imbecile.

"Without this, Ruth would still be upstairs. She needs help, Ben. Not just for the hypothermia, and her lungs. She's messed up."

Ben's face darkened. She knew how hard it was for him to accept that his solid, steadfast wife was losing her grip on reality. But he needed to find some courage.

"Ruth is strong, Ben. She's been there for all of us when we needed her. I don't know what Mum would have done without her."

Her voice tailed off. Ruth had nursed Sonia on the

journey north, had stayed with her all day and all night. She'd done things for Sonia that Jess hadn't been able to face. Jess had spent the last six years coming to terms with the fact that Ruth had been like a daughter to Sonia, and maybe a better one than she had been. She would have to make her peace with it, and teach herself to be less possessive of her mother's memory.

"But right now, she needs help. She's got further to fall than people like you and me. She must be terrified."

"You think she… you think she tried to drown herself?"

Anita sat up straighter in her chair. Jess swallowed. "We can't know that. Not without Ruth here to tell us. But it doesn't matter. What matters is that she gets better."

"Am I such a terrible husband, sis?"

Jess felt a sob leave her lips. She went to Ben's chair and folded herself over him from behind. He was shaking.

"No, Ben. You really aren't."

Ben looked at Anita. "You promise you'll help her. That we'll have access to proper healthcare?"

"Healthcare, and education, and more."

Jess caught herself before she protested at the mention of education. She'd been about to defend the village school where she'd worked for so many years. They'd tried to provide the children with an education. But of course, it hadn't been enough.

"We think there are things we can teach you," she said.

"You do?"

"Yes. We've learned to live off the land. We grow our own food. We share everything. We make good use of limited resources. I imagine the people you work with would benefit from some of that."

"They probably would. But what they need most are places to live."

Ben bristled. Jess nodded. "We'll have to work out the

logistics of that. I'll put together a group, one or two of the villagers who can work with you."

Anita frowned. *She's about to override me,* thought Jess. *She's going to exert her authority.*

Then Anita seemed to catch the look on Jess's face. "Very well," she said. "We'll see what we can do."

55

"It was the night of the riot."

Sarah stood as still as she could, clutching her hands together. The village council surrounded her.

Jess smiled at her. "Go on."

Sarah licked her lips. "I was out checking on my mum." She looked at Dawn, who gave her an encouraging smile. The two of them had spent the whole of yesterday preparing for this. It didn't make it any easier.

"I got back to the flat – Martin's flat – and there was a hole in the window. I found a brick on the floor. It was wrapped in a piece of paper."

"A note?" asked Colin.

Sarah nodded. "Yes."

"What did it say?"

She reached into her pocket, her fingers trembling. She held it out to him.

He took it, his fingers brushing hers.

"Thank you," said Jess. "And then you found something else, posted through the letterbox?"

"Not me."

"Who then?"

"It was Martin who found it." She lowered her voice. "Human excrement." She avoided her mother's eye.

"How do you know it was human?" asked another man. Sanjeev; she knew him from when he had been their neighbour.

"The smell," she said. "Martin works with the chickens, and the pigs. He knows the smell of their – of their waste. It's different."

"It couldn't have been from a dog?"

Jess leaned across the other councillors. "Does that really matter? It's still just as offensive."

Sanjeev shrugged. "Guess so."

Jess turned to Sarah. "And who do you believe did this?"

Sarah glanced at Dawn, who gave her an encouraging nod. "I think it was Sally Angus."

Jess raised an eyebrow. "But it was Mark Palfrey, Sally's fiancé, who accused Martin of attacking him. If the two of them had some sort of disagreement…"

"Martin had never met Mark before. There was no disagreement. Sally wanted him gone. She hated that he'd been part of the group that took us, and that he was here."

Colin cleared his throat. "I don't think it's appropriate to speculate on motive."

"Sorry," Sarah whispered.

"How do you feel about him being here? After what he did to you, and the others?" Colin asked.

Sarah heard a clatter behind her. The council members looked past her, towards the doors. She turned.

Sally Angus was pushing towards Sarah, the doors swinging behind her. Sarah shrank back.

"She's lying!" Sally screeched.

Colin stood up. "Please! You'll have your turn."

Sally pointed at Sarah. Sanjeev and Ben had hurried to Sally and were holding her by the shoulders, keeping her away from Sarah.

"Have you seen the scars on my Mark's face?"

Sarah felt the skin on her arms prickle. She'd seen those scars. She'd seen Sally's nails. She'd put two and two together.

But why?

The doors opened again and Mark walked in. He had less energy than his girlfriend, and less colour in his cheeks.

Colin sighed loudly. "Please, everybody. We'll be taking evidence from one person at a time."

Mark shook his head. "She's lying."

"We've already heard Miss Angus say that."

"Mrs Palfrey!" Sally cried. He's my husband."

Mark stepped towards her. "No, love. We're not married. Not yet." He put an arm around her, resisting Sanjeev who was trying to keep them apart.

"Please," he said. "She's not well."

Jess motioned for everyone to sit down. Someone Sarah didn't know pulled more chairs over, and Mark guided Sally into one. Mark placed himself between Sarah and Sally. It reminded her of the way Sally had put herself between Sarah and Mark when she'd gone to their house.

"This is very irregular," said Colin.

"It's alright," said Jess. "Maybe we can get to the bottom of this more easily, with everyone here. This isn't a court of law, after all."

Sarah felt her muscles relax a little. She trusted Jess. She wasn't so sure about Colin. But he was Sheila's husband; he couldn't be all bad.

She thought of Mark, waiting outside alone. He'd already given his evidence; what had he said?

"Someone ask Martin to come in," said Jess. She ignored Colin's remonstrations.

Martin was given a chair next to Sarah. Sally scowled at him as he sat, but he kept his gaze ahead. He looked nervous.

Sarah grabbed his hand and pulled it into her lap.

"Right," said Jess. She leaned back in her chair. "Mark, carry on."

Mark stood up.

"It's alright. You can stay seated."

Mark sat down. Sally started making small animal-like sounds.

"Thanks," said Mark. He lowered his voice. "I'm sorry, Sally."

The room quietened.

"Please," said Colin. "Say what you barged in here to tell us." Jess frowned at him.

"Sally isn't well. Since the kidnapping."

Sarah felt Martin's hand tense in hers; she held onto it.

"I think she's got PTSD," continued Mark. "She doesn't always know what's real, and what isn't."

"Go on," said Jess, her voice soft.

"She did this to me." Mark pointed at his face. There was a scar running down one cheek. "She was suicidal. I hid all the knives in the house, and she was desperate for me to give them to her. Afterwards, she was insistent that Martin had done it."

Martin's hand loosened in Sarah's. It was damp.

"And the death threat? The parcel through the letter-box?" asked Jess.

"Sally again. I don't think she knew what she was doing." A pause. Sally sniffed. "I'm sorry," continued Mark. "I should have told someone."

"You should," said Colin.

Sally was silent. Sarah risked peering round Mark to look at her. She stared straight ahead, her face stiff. Tears ran down her cheeks.

Poor woman.

"She blamed him," said Mark. "He was a constant reminder. But she went too far. She's not a bad person. She just needs help."

"There's a psychiatrist in Filey," Jess said. "It might be best if the two of you move there."

Mark stood, pushing his chair back. "Is this some kind of punishment?"

"She sent death threats to another member of the village. She made us go through all this."

"She didn't mean to."

Sarah squeezed Martin's hand.

"It's alright," Martin said. Sarah looked at him. How could he be so forgiving?

"We'll move to Filey," Martin continued. He looked at Sarah. "I think it's what we need."

She looked back at him. They'd already discussed this. It was the best way to build their own life together. They wouldn't be far from Dawn, after all. She nodded.

"Very well," said Jess. "But I think Sally owes you an apology, at the very least."

After a moment's silence, Mark spoke. "I'm sorry."

Sally said nothing.

56

"It's hard not being with my boys."

Ruth thought of the conversation she'd had with Sean and Ollie the previous day, when they'd been brought to Filey to visit her. Ollie had wanted to know why they hadn't moved out of the village with her. Sean had put on a brave face, telling his brother to shut up.

But for now, she felt more secure on her own. She'd been given a tiny studio flat in Filey, over a chip shop. She was putting on weight, but the fact that she didn't need to cook or look after anyone was helping her to regain control of her sanity. And using money had been a revelation.

"How do they feel about that?" the therapist asked.

"Confused. They veer from sad to bitter, when I see them. Two hours a week isn't much, but it's all I can take right now."

"That's understandable." The woman leaned forward over the low table. She was white-haired and thin; Ruth wondered if she'd been brought out of retirement for this.

"But I don't feel as if I might hurt them anymore."

"No?"

"No. I can touch them. I can hold them. I just can't be a mother to them. Not properly. Does that make me a bad person?"

"It makes you a person who's unwell. It won't be permanent."

Ruth nodded; it was difficult to believe that, no matter how many times the counsellor said it.

"How are you finding your new accommodation?"

"I like the quiet. No one bothers me."

"Go on."

Ruth frowned. "I don't know. What do you want me to say?"

"Why don't you want people to bother you?"

"That's pretty obvious, I'd have thought."

"I just want to hear it from you."

"I need space. I need time. I want to be back with my boys some day. When I'm… when I'm better."

She swallowed the lump in her throat.

"And your husband?"

Ruth pulled herself inwards. "I don't know."

"Your boys are with him still, in the village."

"Yes."

"You don't have to push yourself too hard, you know. If the thought of being with your boys helps you, then that's great. But it's fine to just focus on yourself right now. Your husband can wait."

Ruth nodded. He would have to. She still didn't know if it would be worth his while, but she knew he would.

"Yes," she said.

The counsellor straightened in her chair. "Our hour is up, I'm afraid. I'll see you next week, same time?"

"Yes. Thanks."

Ruth hated the way their sessions ended so abruptly. It didn't matter where she was in her train of thought, how much she'd managed to come to terms with: the hour was sacred. She shuffled out of the room and down the dark stairs to the street.

The counsellor's office was just two streets away from her own flat, closer to the sea front. She passed a couple in the street. They nodded at her; she had no idea who they were. Was Filey friendly, or was it just that they all knew her, the mad woman from down the coast who'd tried to drown herself?

Jess was standing outside the chippy. It was closed, but the smell of cooking oil still wafted onto the street.

"Hey."

"Hey," replied Ruth. "What brings you to Filey?"

"You do. I came to check up on you."

Ruth didn't believe her sister-in-law. She knew she had business here; she and Anita Chopra were working out the terms of the village being reintegrated into society.

"Come on up."

"Thanks."

Ruth led Jess up the narrow staircase to her flat. She knew it was dirty, with yellow stains on the walls, and that it stank of chips. But she didn't care.

"Coffee?"

Jess pulled a face. "No thanks."

Ruth smiled and put the kettle on. Using electrical appliances again had been a revelation. And she'd rediscovered the addiction to coffee she'd enjoyed in her youth. It probably didn't help her troubled sleep.

"So," she said. "How are Sean and Ollie?"

Jess smiled to herself. "Ollie has lost a tooth."

"Did the tooth fairy come?"

"Of course."

"Ben?"

A blush. "Me. They've been staying with me some of the time."

"Oh." She didn't want to know why. If Ben was falling apart, that was his problem.

"I'll be back soon," she said. "I'm doing well. Making progress."

Jess lowered herself into a chair. "I can see."

Ruth sat next to her. "Thank you."

"What for?"

"You did that deal with Anita Chopra, so I could have help."

"Not just you. Turns out Sally Angus will be visiting the psychiatrist too."

"Oh. Not surprising, I guess."

Ruth sipped at her coffee and stared through the window. Beyond it, seagulls wheeled.

"When I come back," she asked, "will I have to live with Ben?"

"Of course not. You can live wherever you want."

"I want to be with my boys."

"They want to be with you."

"And I might want to be with Ben in time. But not yet. I can't forgive him just yet."

A seagull squawked. "That's fine," said Jess. "He'll understand.

Ruth stiffened. "Will he?"

Jess turned to her. "I'll make sure of it. We'll look after you, Ruth. And I'll look after Ben."

"Thanks. I miss the boys."

"They miss you too."

Ruth smiled. "Give them a hug from me. Tell them Mummy will be home soon."

57

JESS LEFT the coastal path and headed into the village. Seeing Ruth always made her feel better and worse, at the same time. It was good to see that Ruth was getting better, but heartbreaking to see her alone.

She passed her old house on the clifftop. It had a family living in it now: four kids under the age of ten. Sheila would have her work cut out, at least until they started sending all the kids to the school in Filey. She wondered how it would feel, to let all those children leave the village. She had visions of the parents following them up the coast, standing outside the school building until they emerged at the end of the day.

Two boys were ahead of her; young men, maybe. She didn't know them. One wore a grey hoody and the other a sweatshirt with Adidas on the back.

She stopped walking. These were the same boys she'd confronted on her first day as steward, seven months ago. They'd mounted the climbing frame, waving a banner between them: *go home skum*.

Did they live here now? Or were they causing trouble?

"Boys!" she called to them. They turned.

She felt herself deflate. These boys were too young. They were no more than thirteen, and their faces were smoother. They exchanged glances then ran off between the houses, calling to each other good-naturedly.

She would have to get used to this. The village wasn't what it was. They had more people to deal with, more mouths to feed. But they had help. There was a doctor in Filey now, and a school. And there was money. The men who went to Hull to work on the earthworks, Sam still among them, earned a proper wage. Sam was a team leader, earning five times what he had. It wouldn't be long before they were expected to have bank accounts.

She stopped walking. Sometimes, the grief hit her like a bullet. She thought of Zack, strolling through the village with her, towards his family's house. On the night they'd told them about the engagement. They'd held hands, swinging their arms between them. For the first time in over six years, her mind had been on the future.

What now? Stay here, try to steer this village into an uncertain future? Or move to Filey, where she wouldn't be reminded of her fiancé every time she turned a corner?

Filey was still a ghost town; that much had been clear from this last visit. The area where the gas explosion had taken place was hazardous, the wreckage of buildings being hastily blocked off by the authorities. But some of the villagers had chosen to live in the town. And it was giving the place a new lease of life. Toni and Roisin were among them. She hoped they'd be happy there. Toni had told her Flo was coming round to the idea gradually, but it could take a while.

Ben emerged from his house as she passed.

"Did you see her?"

She stopped walking, irritated. "Yes."

Be patient, Ben. That was what she'd said to him, on the day she'd done the deal with Anita Chopra.

"How was she?"

"She was drinking coffee."

"Coffee?"

"She offered me a mug. I said no."

"Is that a good thing?"

"I guess so." She didn't tell him Ruth wanted to come back to her boys, but not her husband. *Cross that bridge when we come to it.*

He said nothing, but fell into step beside her. They were heading for the village hall, for the council's final meeting. The group was to be disbanded. Next month the villagers would be voting in local elections. She tried to muster enthusiasm for electing members of Filey town council, but couldn't bring herself to. Maybe that would change. Maybe she would stand herself.

"Give her time, Ben."

"I want her back."

"If you push her, you'll lose her for good."

"Put a good word in for me, will you? I haven't seen her for weeks."

She grabbed his hand. "I know, bruv. I'll do what I can."

They pushed open the door to the village hall and she steeled herself for her last meeting as steward.

Read more about the villagers in *Underwater*, the prequel stories to this book. Get your FREE copy at rachelmclean.com/underwater.

READ THE PREQUEL

Find out how the villagers arrived in Yorkshire in the prequel, *Underwater*.

'Hurricane Victoria, they called it. Such a British name. So full of history, and patriotism, and shades of Empire.'

Little did they know it would devastate London and send an exodus of refugees north.

In this companion set of prequel stories to the Village trilogy, discover how the Dyer family are forced to leave London as it descends into chaos. **Will they reach Leeds and their eventual coast destination safely?**

Get your FREE copy of *Underwater* at rachelmclean.com/underwater.

Happy reading!

Rachel McLean

A HOUSE DIVIDED, PART 1 OF THE DIVISION BELL TRILOGY

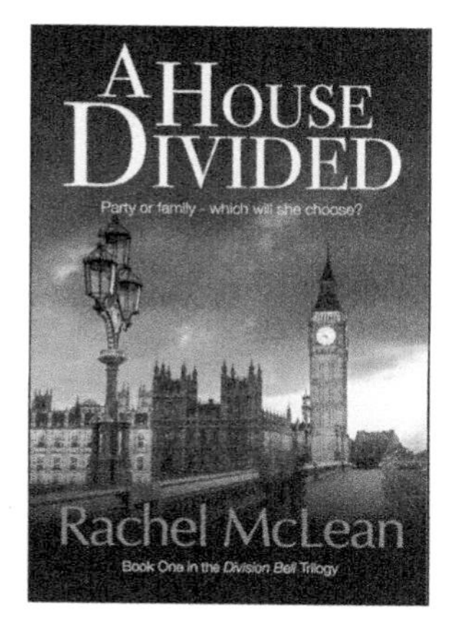

Jennifer Sinclair is many things: loyal government minister, loving wife and devoted mother.

But when a terror attack threatens her family, her world is turned upside down. When the government she has served targets her Muslim husband and sons, her loyalties are tested. And when her family is about to be torn apart, she must take drastic action to protect them.

A House Divided is a tense and timely thriller about political extremism and divided loyalties, and their impact on one woman.

Available from Amazon in paperback and e-book.

Printed in Great Britain
by Amazon